Chris Defonseka

Two-Component Polyurethane Systems

I0036715

Also of interest

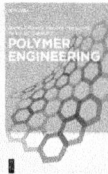

Polymer Engineering
Tylkowski, Wieszczycka, Jastrzab (Eds.), 2017
ISBN 978-3-11-046828-1, e-ISBN 978-3-11-046974-5

Water-Blown Cellular Polymers
A Practical Guide
Defonseka, 2019
ISBN 978-3-11-063950-6, e-ISBN 978-3-11-064312-1

Polymeric Composites with Rice Hulls
An Introduction
Defonseka, 2019
ISBN 978-3-11-063968-1, e-ISBN 978-3-11-064320-6

Flexible Polyurethane Foams.
A Practical Guide
Defonseka, 2019
ISBN 978-3-11-063958-2, e-ISBN 978-3-11-064318-3

e-Polymers.
Editor-in-Chief: Seema Agarwal
ISSN 2197-4586
e-ISSN 1618-7229

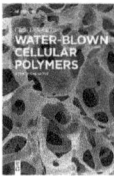

Chris Defonseka

Two-Component Polyurethane Systems

Innovative Processing Methods

DE GRUYTER

Author
Chris Defonseka
Toronto
Canada
defonsekachris@rogers.com

ISBN 978-3-11-063957-5
e-ISBN (PDF) 978-3-11-064316-9
e-ISBN (EPUB) 978-3-11-063979-7

Library of Congress Control Number: 2018964971

Bibliographic information published by the Deutsche Nationalbibliothek
The Deutsche Nationalbibliothek lists this publication in the Deutsche Nationalbibliografie;
detailed bibliographic data are available on the Internet at http://dnb.dnb.de.

© 2019 Walter de Gruyter GmbH, Berlin/Boston
Typesetting: Integra Software Services Pvt. Ltd.
Printing and binding: CPI books GmbH, Leck
Cover image: Steve McAlister/gettyimages

www.degruyter.com

Preface

Polyurethanes are an important branch of plastics belonging to the *thermosetting group* and have seen rapid advances over the years with speciality grades, in addition to the primary applications of comfort. There are many types and grades of polyurethanes ranging from soft to rigid and then again to thermoplastic polyurethanes to microcellular foams. These polyurethanes have become the preferred material over others, for manufacturing essential products for comfort, industrial and consumer products, automobile components, building construction applications, space travel, footwear and many others.

In the earlier days, processing of polyurethanes involved complex methodologies based on mixtures of many different chemical components but because of constant research and development two-component systems began to appear and today these easy to process systems are available for most grades of polyurethanes. While the large volume producers may still prefer the 'multi' systems, the two-component systems will provide the smaller producers, prototype and speciality moulders with many advantages in ease of handling, simpler processing and meaningful cost savings. They will also have the advantage of getting custom-formulated systems from professional suppliers.

Although polyurethanes are based on complex polymer chemistry, the presentations in this book make it easy to understand the basics of polyurethanes, the functions of each component, guidelines on formulating to achieve desired end properties, innovative processing and moulding methods to make products as needed by the vast spectrum of important end applications. The author also presents mould designs, selection of machinery and recommended solutions for post-moulded defects. Some of the main industries that will greatly benefit from the availability of these two-component systems are automotive, building construction, footwear, manufacturing, coating and adhesive sectors.

The author's presentation of a complete plant designed and built by him to produce large size foam blocks based on two-component systems for *flexible, high resilience* and *viscoelastic foams* should be of great interest to small volume foam producers and to entrepreneurs, as they can fabricate similar processing machinery themselves to avoid investing in standard machinery and equipment, which are capital intensive. Also presented is a case study carried out by the author to reduce foam waste in a large plant from 33% to 14%, how to improve foam quality and process efficiency and enhance profitability.

This book deals with all aspects of most two-component polyurethane systems based on important applications. It is hoped that the readers of this book will benefit from the information and processing innovations provided by the in-depth knowledge and hands-on experience of the author, spanning over 45 years in the field of polyurethanes and still active, both locally and internationally.

https://doi.org/10.1515/9783110643169-201

Contents

1 Introduction to polyurethanes

1.1 An overview

Polyurethanes(PUs) are an important branch of plastics belonging to the *thermo-setting group* unlike other polymers such as polyethylene (PE), polystyrene (PS), polypropylene (PP), polyvinyl chloride and many others which belong to the *thermoforming group*, meaning these polymers can be re-used for production. PU materials having a wide range of versatile properties, including pleasing aesthetic values, find many essential applications in important sectors such as comfort, automobile, building construction, footwear, coating, space travel, industrial, consumer products, packaging, adhesives and others.

Unlike the thermoforming polymers which are derived from their monomers, PUs do not have a basic monomer as such to start with and the urethane monomer is produced by reacting a polyol (an alcohol with more than two reactive hydroxyl groups per molecule) and a diisocyanate or polymeric isocyanate in the presence of suitable catalysts and additives with water as the primary blowing agent to form a foam. Since a wide variety of diisocyanates and a wide range of polyols can be used to produce PUs, a broad spectrum of materials can be produced to meet the needs of specific applications.

PUs can be broadly categorised as follows: *flexible, semi-rigid, rigid, microcellular, viscoelastic or thermoplastic urethanes*. When considering comfort applications such as mattresses, cushions, sheets, slab-moulded automobile and aircraft seats and viscoelastic products and the medical applications, one may conclude that the flexible foams have the highest number of demand.

Viscoelastic foams, also known as memory foams, are PU foams with special ultra-comfort properties. Standard flexible PU foams used for comfort applications are two-dimensional, while viscoelastic foams with four-dimensional properties such as *density, hardness, temperature* and *time* that can be varied, gives much superior comfort. When they first arrived on the market, it was called a 'miracle foam', their speciality properties being ultra-comfort and great therapeutic values, although they were heavier and also cost more than standard PU foams. Viscoelastic foams are generally graded by their density and thicknesses and foams selected by most people for a mattress will range from about 4 inches (10.2 cms) to 8 inches (20.3 cms), although a thinner topper for an existing bed, is also an option. The current availability of custom-ordered two-component systems for their productions is a great boost for these foam producers.

Years back, the production of PU foams involved the mixing of multiple components needing careful formulating and weighing each chemical accurately which had to be precise to prevent foam wastes. Now, most PU systems, except for the very large volume foam producers are available as two components and producers have the options of using standard systems available in the market or custom made to

https://doi.org/10.1515/9783110643169-001

meet individual end application needs. However, they are not without disadvantages, as each system can produce a foam of one density only with predetermined properties. More advanced systems are also available, where the densities can be adjusted by variations in the mixing ratios.

Large volume PU foam producers with continuous systems will opt for purchases of raw material components in bulk, storing them in large tanks connected to a central mixing head and draw from them on pre-set quantities. These systems will allow a foam producer products of varying densities and properties but the foam wastes can be higher. However, for other applications like coatings, mouldings, integral skin moulding, in-situ building insulation, appliance insulation, sound proofing, footwear mouldings and others, the use of two-component systems would be more practical and effective with lesser costs.

The emergence of two-component PU systems, although slow at the beginning, has been a great boost in foaming. Because of rapid advances in technology, most of these two-component systems can be 'hand-mixed' in simple proportions of component A and component B or processed using simple mixing machines. Some are even available in small cans, containers or small packs for instant processing and applications.

The presentation of identifying and use of ideal biomass fillers and methodologies for increasing thermal conductivity in PU foams should be of interest to foam producers and in particular to the automotive industry. Detailed information of both solvent borne and water borne two-component PU systems which have emerged in the last few years, also should be of interest.

To impart valuable knowledge and encouragement to readers, small foam producers and would-be entrepreneurs to realise the exciting possibilities of two-component systems, the author presents a complete cost-effective design of a foaming system and a simple cutting machine suitable for making large blocks of *flexible, high resilience* and *viscoelastic foams* using two-component PU systems, based on an actual project designed, fabricated and implemented by the author. Also presented are information on formulations, processing techniques, a two-component raw material system for large foam blocks, mould designs, handling and safety factors. The highlighting of specialty two-component systems developed and made available by well-known companies, along with lists of key raw material suppliers and machinery suppliers should provide useful additional information.

1.2 Chemistry of polyurethanes in brief

The polymeric material known as PU forms a family of polymers which are essentially different from most of the other plastics in that there is no urethane monomer and this polymer is invariably created during the manufacture of a PU product.

PUs are made by the exothermic (heat emitting) reactions between alcohols with two or more reactive hydroxyl (OH) groups per molecule (diols, triols and polyols) and isocyanates that have more than one reactive isocyanate group (NCO) per molecule (diisocyanates and polyisocyanates). For example, when a diisocyanate reacts with a polyol, the group formed by the reaction between the two molecules is known as the 'urethane linkage'. This is the essential part of a PU molecule.

The physical properties, as well as the chemical structure of a PU depends on the structures of the original reactants. The characteristics of the polyol/polyols used such as relative mass, the number of reactive functional groups per molecule and the molecular structures will influence the final properties of the polymer. There is a fundamental difference between the manufacture of most PU products and the manufacture of many other plastics products.

Polymers such as PE, PP, PS and others are produced in chemical plants in the form of granules, powders or others. Products are made from these basic raw materials by heating, shaping under pressure and cooling by different processing methods. The properties of these end products will almost completely depend on the original polymer. On the other hand, raw material components for PUs are liquids and usually made directly into products using mixing processes, with options for adjusting for desired properties at the time of formulating.

1.3 Some important polyurethanes

PUs are very versatile materials and can be made with textures from soft, rigid to elastic with varying densities and properties. With the advent of waterborne systems and PU composites made up of non-traditional fillers like biomass flours/powders, bamboo fibres and so on, applications are increasing rapidly. Some of the important areas of foams are provided in the following sections:

1.3.1 Flexible foams

Flexible PU foams account for about 30% by volume being used for bedding, furniture and in the automotive and aircraft industries. Viscoelastic foams, which can also be classified as flexible foams are largely used for bedding. Flexible foams can be made in almost any variety of firmness. They are durable, supportive and comfortable, with viscoelastic foams providing ultra-comfort.

1.3.2 Rigid polyurethane foams

Rigid PU and polyisocyanurate foams create one of the most popular, energy-efficient temperature and lower noise levels in homes and commercial properties, builders use rigid PU and polyisocyanurate foams. These foams are effective insulation materials that can be used in roof and wall insulations, insulated windows, doors and as air barrier sealants.

1.3.3 Coatings, adhesives, sealants and elastomers

The use of PUs in coatings, adhesives, sealants and elastomers offer a broad and growing spectrum of applications and benefits. PU coatings can enhance a product's appearance and also lengthen its lifespan. PU adhesives can provide strong bonding advantages, while PU sealants provide tighter seals. PU elastomers can be moulded into any shape, are lighter than metal, offer superior stress recovery and can be resistant to many environmental factors.

1.3.4 Thermoplastic polyurethane

Thermoplastic polyurethanes (TPU) offer a myriad of combinations of physical property and processing applications. They are highly elastic, flexible and resistant to abrasion, impact and weather. TPUs can be coloured or fabricated in a wide variety of methods with product durability as a special feature. TPU is an elastomer that is fully thermoplastic. Like all thermoplastic elastomers, TPUs are melt-processable. In addition, they can be processed on extrusion, injection, blow moulding and compression moulding equipment. They can also be vacuum-formed or solution-coated and well suited for a variety of fabrication methodologies. They have many uses in applications in the building construction, automobile and footwear industries.

1.3.5 Reaction injection moulding

Car bumpers, electrical housing panels and computer telecommunication equipment enclosures are some of the parts that are produced with PUs using reaction injection moulding (RIM) techniques. RIM process produces parts that are not achievable using standard injection moulding processes. In addition to high strength and low weight, PU RIM parts can exhibit heat resistance, thermal insulation, dimensional stability and a high level of dynamic properties. Automotive, construction, appliances, furniture, recreation and sporting goods are a few of the markets using RIM technologies.

1.3.6 Binders

PU binders are used to adhere numerous types of particles and fibres to each other. Their primary areas of use are in the manufacturing of wood panels, rubber or elastomeric flooring surfaces and sand casting for the foundry industry. PU binders are widely used in the manufacture of oriented strand board (OSB). These wood panels are used in structural sheathing and flooring, pre-fabricated housing and so on. Re-bonded foam underlay uses PU binders to adhere scrap foam pieces, which are often flexible PU foams.

1.3.7 Waterborne polyurethane dispersions

Waterborne polyurethane dispersions (PUDs) are coatings and adhesives that use water as the primary solvent. With increasing environmental concerns and stricter regulations on the amount of volatile organic compounds (VOCs) that can be emitted into the atmosphere, PUDs are being used in more industrial and commercial applications.

1.4 Some polyurethane applications

1.4.1 Apparel

When researchers discovered that PUs can be made into fine threads, they were combined with other polymers to make lightweight, stretchable materials ideal for garments. Over the years, PUs have been improved into what is known as spandex fibres. Because of today's constant advances in PUs and processing techniques, manufacturers can make a broad range of apparel for garments, sports clothes and variety of accessories.

1.4.2 Appliances

PUs are an important component in major appliances that are used by consumers every day. The most common are rigid foams as insulation for refrigerators and freezer thermal insulation systems. These rigid foams are an essential and cost-effective material with easy-flow properties meeting required energy ratings in consumer appliances. The good thermal insulating properties of rigid PU foams result from the combination of fine, closed-cell foam structures and cell gases that resist heat transfer.

1.4.3 Automotive

PUs are widely used in many car and transport vehicles. In addition to the foams that are used for making comfortable seats, bumpers, interior ceilings, car panels, spoilers, doors and other applications all use PUs in various forms. PUs also enable manufacturers to provide drivers and passengers more significant automobile 'mileage' by reducing weight and thus increasing fuel economy, comfort, corrosion resistance, insulation and sound absorption.

1.4.4 Building construction

Today's homes and buildings demand high-performance materials that are strong, lightweight and easy to install with good durability. PUs help conserve natural resources and helps the environment by reducing energy use. With its excellent strength-to-weight ratio, insulation properties, durability, versatility and PU are frequently used in building construction applications. Both the affordability and the comfort they provide homeowners have made PUs a part of homes everywhere. Carpet underlay, insulation and sound proofing play a big role. PU building materials add design flexibility to new homes and remodeling projects. Foam-core panels offer a wide variety of colours and profiles for walls and roofs, while foam-core entry doors, separation panels and garage doors are available in different finishes and styles. The availability of PU composite materials also plays a major role.

1.4.5 Electronics

Non-foam PUs, often called 'potting compounds' are frequently used in electrical and electronics industries to encapsulate, seal and insulate fragile, pressure-sensitive microelectronic components, under water cables and printed circuit boards. PU potting compounds are specially formulated by developers to meet a diverse range of physical, thermal and electrical properties. They can protect electronics by providing excellent dielectric and adhesive properties, as well as exceptional solvent, water and extreme temperature resistance.

1.4.6 Flooring

PUs are also used to coat floors, from wood and parquet to cement. These protective finishes are resistant to abrasions and solvents, easy to apply for cleaning and maintaining. High gloss non-scratch coatings are speciality applications. These special coatings can be used to re-finish for old floors and surfaces to look new again.

1.4.7 Furnishings

PU foams are the best for comfort applications like bedding and furniture. Artificial leather made with PU coatings are very soft (softer than PVC coatings) and can be made to imitate natural leather with exotic surface designs to improve their aesthetic values. PU foams and coatings are widely used in boats, while PU epoxy resins seals boat hulls from water, corrosion and the weather. Boaters, mobile trailer homes and others can also have home comforts due to the many uses of PUs. Rigid PU foams insulates boats from noise and severe temperature extremes and increases load-bearing capacity with minimum weight.

1.4.8 Medical

PUs are commonly used in a number of medical applications like catheter and general purpose tubing, hospital bedding, surgical drapes, wound dressings and a variety of moulded devices. Medical implants are common uses providing longevity and toughness. Viscoelastic foams are replacing standard flexible foam bedding because of their therapeutic values, especially because of prevention of bed sores. These foams are also used for wheelchair seats and post-operative devices.

1.4.9 Packaging

PU packaging can provide form-fitting cushioning that uniquely holds a product firmly in place during transit. Foam-in-place (in-situ foaming) is a two-component rigid PU foam which can be foamed around any product inside a cardboard box for transport of fragile parts such as electronic and medical equipment, delicate glassware, ceramics and many others. These foams 'cure' instantly and can be carried out anywhere. These packaging are specially used for long haul transport or where packs can possibly undergo stress by movement or high loads.

1.4.10 Speciality mouldings

The choice of NASA scientists was viscoelastic PU foam cushioning for countering the g-forces encountered during space travel. The four-dimensional versatility of these special foams which can be varied, greatly helps. Probably, the largest volumes produced next to bedding and furniture applications are automobile and aircraft seats. These are special mouldings using flexible or high-resilience Polyurethane (PUR) foams, with choices of hot or cold curing methods. Self-skinning finishes with aesthetic values are a big advantage. Because of therapeutic values most hospitals and medical centres have been phasing out the old bedding and replacing with visco-elastic foam slabs or the specially moulded ones. In the case of the former, combinations of flexible, viscoelastic foams or other, can be combined with viscoelastic foam being the top layer for cost-effectiveness. Here, recycled foam wastes in the form of slabs would be a good base.

Bibliography

1. American Chemistry Council – article: 'Introduction to Polyurethanes'. www.americanchemistry.com.
2. Polyurethanes: essentialchemicalindustry.org/polymers/polyurethane.html.
3. Defonseka, Chris – author. Practical Guide to Flexible Polyurethane Foams – ISBN 978-1-84735-974-2, 2013.

2 Types of polyurethanes

The material structure of polyurethanes(PUs) can be basically categorised as open-cell and closed-cell, with flexible foams having open-cells and rigid foams having closed-cells. Open-cell foam is soft like a foam cushion that is used for comfort or a viscoelastic foam mattress used for ultra-comfort. The formation of open-cells takes place during the foaming process with the cell walls or the surfaces of the bubbles being broken and air filling all the spaces in the material. This makes the foam soft or weak as if it were made of broken balloons, with the size of the cells pre-designed and controlled with additives. No doubt, that a flexible PU may have a few non-standard cell sizes or even a small area or areas consisting of non-conforming cell sizes but with modern processing technology, these possibilities have been eliminated to a negligible degree. The insulation value of these foams is related to the insulation value of the calm air inside the matrix of the broken cells and also the foam density.

Closed-cell foams will have varying degrees of hardness, depending on the density. A normal closed-cell insulation or PU flotation is between 2 and 3 pounds per cubic foot. Most of the cells or bubbles in the foam are not broken and they resemble inflated balloons piled together in a compact configuration. This makes the foam strong or rigid because the bubbles are strong enough to withstand a lot of pressure. The cells are full of gas, selected to make the insulation value of the foam material as high as possible.

The advantages of the closed-cell foam compared to open-cell foam include its strength, R-value and much greater resistance to the leakage of air or water vapour, when used as an insulation material. The disadvantage of closed-cell foams is that it is more dense, requiring more material and therefore, more expensive. Even though it has a higher R-value, the cost per R is still higher than open-cell foam. The choice of open-cell or closed-cell foam will largely depend on the end application. Both types of foam are commonly used in most building applications.

There are many different types of PUs and the following are selected for presentation based on the availability of two-component systems for processing them.

2.1 Flexible foams

The 'flexible' family consists of a wide spectrum of open-cell foams: conventional, high-resilience foams and viscoelastic foams that are mainly used for comfort applications.

https://doi.org/10.1515/9783110643169-002

2.1.1 Conventional flexible foams

These flexible PU foams first came on the market in the 1950s. They are called PU foams, PUR foams or Flexible Polyurethane Foam (FPF) and their acceptance was immediate as a superior material to traditional ones such as cotton, rubber and fibre. Compared to rubber latex foam, these flexible PUR foams are superior in many ways. They have greater tear strength, better resistance to oxidation, aging as well as fire and their open-cell structures enable them to absorb and emit body heat. These foams are much lighter than rubber latex foams that can be produced in densities ranging from 12 to 50 kg/cu.m. For two-component systems' manufacture of these foams, processes include manual, semi-automatic or discontinuous (single block) processes. The two important basic properties of conventional flexible foams for marketing are as follows: density and indentation force deflection (IFD) factor which should be ≥2.0 on a 65% : 25% indentation ratio.

2.1.2 High-resilience foams

High-resilience foams called HR foams and also known as MR (maximum resilience) foams are open-cell, flexible PU foams that have less uniform (more random) cell structures that helps add support, comfort, resilience and bounce. These foams have a high support factor and greater surface resilience than even viscoelastic foams. HR foams have a very fast recovery and bounce back to its original shape immediately after compression. HR foams have properties closest to rubber latex foams which are also used for cushioning and bedding but HR foams give better comfort and last much longer and in most grades have a resilience factor greater than 45°. HR foams can be produced in any colour but the normal standard colour is grey. Table 2.1, as adapted by

Table 2.1: Grades and specifications of HR foams.

Grade	Density (kg/cu.m)	Hardness 'n' @40% compression	Resilience	Standards	Typical uses
28 HR	28	50–70	55% min	ASTMD3574	Furniture Cushions
32 HR	32	80–100	50% min	ASTMD3574	Mattresses Furniture Cushions
35 HR	35	170–230	45% min	ASTMD3574	Firm seating
60 HR	60	130–170	48% min	ASTMD3574	Commercial Transport Seating

data given by Sheela Foam Limited (India), shows grades and typical specifications of HR foams in general.

Some of the markets for HR foams are as follows:

– Furniture
– Bedding and overlays
– Speciality seating
– Aerospace
– Automotive
– Consumer products
– Design
– Hospitality and hotels
– Medical comfort applications
– Packaging
– Apparel padding
– Military
– Toys

2.1.3 Viscoelastic foams

Viscoelastic foams or memory foams when they first came on the market created a sensation and people called it a 'miracle foam'. A viscoelastic foam is an open-cell flexible PUR foam specially formulated to increase its density and viscosity, and has low resilience. While conventional foams are two-dimensional, viscoelastic foams are four dimensional with special properties such as density, time, hardness and temperature. Densities for viscoelastic foams range from 72 to 85 kg/cu.m with IFDs between 12 and 16. Some manufacturers may produce lower density foams but the market will grade them as low-end foams.

These speciality foams are mainly used as mattresses or mattress toppers and work by conforming to normal heat, weight and shape of a body. They adjust to accommodate the various pressure points in the body by 'accepting' the body, cushioning it and allowing it to sink into the foam without resistance and thus creating a sensation of floating on a cloud. When the body is removed, the foam slowly 'recovers' to its original shape. All four dimensions can be adjusted to suit the end properties as desired during formulating.

These foams are much heavier than conventional or high-resilience foams and when making large blocks, it is recommended to limit the foaming height to 28 inches (70 cm) instead of the normal 42 inches (105 cm) for easy post-cure handling of the blocks. Also the final-cure phase of these blocks will be around 48 hours instead of the standard 24 hours for conventional foams.

2.2 Spray foams

Many types and grades of polyurethane spray foams (PSF) are available in the market and their main applications are insulation of buildings. For ease of application, these versatile foams are available as two-component systems in small size cans (small applications) to large containers that are complete with spray gun attachments. Basic systems are mixing and reacting chemicals to create foams. These foams can be either be open-cell or closed-cell foams varying from low density to high density. While low pressure systems may be used for small applications, high pressure systems are generally used for insulation of large areas and buildings.

The reacting materials react very fast, quickly expanding to form a foam that insulates, air seals and also provides a moisture barrier. PSFs are known to resist heat transfer extremely well and it offers highly effective solutions in preventing unwanted air infiltration through cracks, seams and joints. Professionals use both low pressure and high pressure systems, depending on the end application. Whether retro-fitting a home or building a new one, builders and homeowners have the view that PSF insulation is a great way to save on energy costs and improve comfort. These cellular polymeric materials flow easily into any crevices, however small they are and provides full insulation by fully filling all voids. There are several major differences between the open-cell and closed-cell foam applications, leading to advantages and disadvantages, depending on the end application requirements. Table 2.2 highlights some of these advantages and disadvantages.

Table 2.2: Typical advantages and disadvantages.

Closed-cell	Open-cell
R-value greater than 6.0 per inch R = thermal resistance	R-value around 3.5 per inch
Low moisture vapour permeability	Higher moisture vapour permeability
Air barrier	Air barrier at full wall thickness
Higher strength and rigidity	Lower strength and rigidity
Resists water	Not recommended for applications in direct contact with water
Medium density (1.75–2.25 lbs./cu.ft.) (28–36 kg/cu.m)	Low density (0.4–1.2 lbs./cu.ft) (6.4–19.2 kg/cu.m)
Absorbs sound	High sound absorption

2.2.1 High-pressure two-component PU spray foams

Three of the common high-pressure systems are as follows:
– Open-cell 0.5 lb. low density foam

- Closed-cell 2.0 lb. medium density foam
- Closed-cell 3.0 lb. high density foam

These three types of spray foams used in building insulation and weatherisation, the high building construction or renovations. It is always advisable for trained professionals to handle high-pressure spray foam applications. These insulating systems operate at 1,000 psi or higher. The two-component chemicals are typically brought to the job site in 55 gallon drums with a spray rig, air supply, high-pressure rated hoses and a spray gun completing the system.

A high-pressure foam rig can quickly deliver continuous foam to a large area, insulating walls, beams and roofs. As the foam expands and hardens, the foam seals small gaps and crack that allow air leaks. These foams will also provide high resistance to heat transfer. PSFs applied with a high-pressure system adheres tightly to the entire structure without sagging or separation from its position. They provide complete seals against air, moisture, heat and insects. These insulations are unique in that they keep homes and buildings cool in summer and warm in winter and helps to reduce heating and cooling bills by controlling the amounts of energy used within.

2.3 Semi-rigid foams

Semi-rigid PUR foams are in great demand for applications, especially in the aircraft, automobile and transport industries. Most in demand probably is for open-cell integral skin moulded products, where a smooth, tough outer skin protects the cellular foam core inside. In automobile applications like seats, armrests, dashboard, panels and others, these semi-rigid foam systems are used for foam backing films, skins or leather covers. Depending on the end application needs to meet regulations, these systems can be tailor-made such as Elastoflex E made by BASF.

In the automobile sector, noise and vibration damping can provide high driving comfort vehicle interiors. Semi-rigid foams with its excellent damping properties can absorb and distribute these impacts effectively. Because of their open-cell structures, they have special acoustic properties which will eliminate rattling and humming. In addition, with fogging values below 0.5 mg, emissions in the passenger compartments will be very low.

A new state-of-the art 'honeycomb' technology with the use of clever combinations of paper honeycombs and PUR foams can manufacture very light vehicle interior components that are capable of good stress, sound and vibration absorption and used in areas where enhanced properties are needed. These will, in addition, contribute to vehicle weight reduction and reduced fuel consumption.

2.4 Rigid foams

Rigid foams are closed-cell foams and builders turn to these PU foams and polyisocyanurate (polyiso) foams, which are today's one of the most effective insulation materials available for roof, wall insulation, insulated windows and doors and as air barrier sealants. They have high cross-linking density with good heat stability, high compression strength and excellent insulation properties.

The growth of the rigid foam market is driven by the increasing use of rigid foams in the building and construction and appliances industry for insulation and energy saving and in the automobile industry for purposes of weight reduction. These versatile foams are also widely used in the packaging industry owing to its light weight and durability.

2.4.1 Energy efficiency

- PU and polyiso foams probably have one of the highest insulating R-values per inch of all commercially available products today. With typical R-values in the range of 5.5–8.0 per inch, it is possible to have thinner walls and lower profile roofs, while maximising efficiency, increasing space utilisation and reducing costs.
- Increasing the roof insulation by one inch or more, above the accepted standards, can reduce energy costs savings substantially.
- Entry doors having a rigid PU foam core will help inhibit sound and add insulation value that will further reduce heating and cooling energy needs.
- PU foam sealants, applied on-site, will expand to fill energy-wasting, air-infiltrating gaps around window frames, plumbing pipes and electrical outlets.
- 'Reflective' plastic coverings over PU foam insulated roofs, bounce sunlight and radiant heat away from a building, helping the structure stay cool and reducing energy use for air-conditioning.

2.4.2 High performance

Rigid PU and polyiso foams are made with a remarkably strong, yet lightweight, low-density structures that are both dimensionally stable and moisture-resistant with low vapour transmission. This special combination of properties allows manufacturers to design thermal insulating products that are self-supporting, can be combined with a wide range of substrates, while requiring no additional adhesives and perform well as exterior weather and moisture barriers.

High-performance rigid PU insulation can be applied by spraying onto various UV degradation and re-coated to extend longer performance periods. These rigid foams when properly installed are not affected by oil-based waterproofing.

Both PU and polyiso foams are widely used in fabrication of steel-faced building panels for various categories of commercial building constructions. General applications range from cold storage warehouses, buildings used for food and beverage industries to high-tech offices, medical buildings, airports and even manufacturing facilities. Foams can be easily bonded to metal skins which are very useful property when large high-strength buildings are constructed. The lightweight and high insulation value make these products ideal for renovating exteriors of older buildings because typically the existing cladding can be left in place and the existing structure does not have to be reinforced.

2.5 Microcellular polyurethane foams

The concept or technology of microcellular polyurethane foams (MPU) possibly is an interesting extension of the cross-linked polyethylene foams. Nonetheless, the advantages of micro-celled foam has spurred many commercial uses and have spread quickly to most areas of applications, especially for the footwear, automotive and other industries. Prominent among these micro foams were the polyethylene, polyvinyl foams and later on MPU, probably the most versatile and widely used micro foam.

Conventional cellular foams are chemically blown with non-volatile blowing agents as compared to microcellular foams (MCF) blown with volatile carbon dioxide or nitrogen. The latter process is aggressive, lowly nucleated with larger expansion. MCFs contain around 108 cells per cu.cm. and cellular foams around 104–106 cells per cu.cm.

The aggressive expansion in microcellular production makes the polymeric strength very critical in maintaining cell strength, more so in continuous foaming where cell coalescence can take place. Thus, thick cell walls are very necessary and that lays the expansion limit to about 10 times, whereas the cellular foaming can achieve over 50 times expansion. The end uses for cellular and microcellular applications are different, with cellular foams probably having a larger volume demand in the market. However, when expansion reduces to 30–70% weight reduction, about two to three times expansion, quite a few polymers qualify for foaming. This becomes a great opportunity for engineered polymers, where material savings is rather substantial. Nylon, ABS, PC and filled-PP are good examples.

Foaming is a phase separation phenomenon governed by thermodynamically driven kinetics. A common practice is to establish a positive superheat or supersaturation, that volatile phase which tends to conglomerate into spherical gas bubbles. In general, saturation with gas and then applying vacuum or heat or both, to induce thermodynamic instability where bubbles appear. Traditional cellular foaming technology controls nucleation via a nucleating agent, whereas in microcellular foaming by super critical carbon dioxide, which plays a dual role as a blowing agent and nucleating agent. As a general rule, the less the expansion, the finer the cells. Fewer than 10 times expansion will give 10 microns or under. At thirty times expansion, the cells are in the hundred microns.

The unique features of MCF are very fine high-density cell sizes. Since the cells are very fine, without careful attention MCF may be seen as a plastic material, rather than a cellular product. Conventional polymeric foam is known for its high performance/weight ratio, which increases as cell size decreases and cell integrity improves.

Microcellular products have many applications in industrial and other fields. For information of the reader, the following are presented:

2.5.1 Microcellular polyurethane elastomer

Cellasto is a microcellular polyurethane elastomer (MPU) sold as a finished product mainly for the automotive industry. Under the brand name – Cellasto – BASF develops, produces and distributes components that enhance driving comfort. Cellasto is not only more robust and durable, but also has superior physical properties when it comes to minimising noise from the engine and isolating vibrations from the chassis, shock absorbers and struts, which are important challenges in the auto industry. The MCF saves space and weight because of its excellent volume compressibility.

Some of the benefits are as follows:
- Low compression set
- High abrasion resistance
- Ozone resistance
- Cadmium-free
- Very good static and dynamic long-term behaviour
- High-volume compressibility with minimum lateral expansion
- Resistance to oils, greases and other aliphatic hydrocarbons

2.5.2 Microcellular urethanes

PORON microcellular urethanes specially made by Rogers Corporation for gasketing and sealing provide solutions for the following:
- Communications
- Electronics
- Automotive
- Industrial equipment and devices

Some of the main benefits are as follows:
- Excellent compression set resistance with long-term durability
- Low outgassing and non-fogging. Does not contain any plasticisers
- Inherently flame retardant without the use of additives
- Engineered urethane formulations offers a wide modulus range 2–90 psi @25% deflection
- Good chemical resistance
- Easy to fabricate. Clean die-cuts and works with a broad range of adhesives.
- Many grades with thicknesses from 0.012 inches to 0.500 inches (0.43 mm– 12.7 mm)

2.6 Thermoplastic polyurethane

Thermoplastic polyurethane (TPU) is an elastomer that is fully thermoplastic. Like all thermoplastic elastomers, TPU is elastic and melt-processable. It can be processed by extrusion, injection moulding, blow moulding or compression moulding. Fabrication methodologies will include vacuum forming and solution-coating as well for a wide variety of applications. To improve aesthetic values it can be coloured but more important is that it can provide a considerable number of physical property combinations, making it an extremely adaptable material for dozens of uses.

This versatility is partly because TPU is a linear segmented block copolymer composed of hard and soft segments. The hard segment can be either aromatic or aliphatic. Aromatic TPUs are based on isocyanates such MDI, while aliphatic TPUs are based on isocyanates like H12 MDI. When these isocyanates are combined with short-chain diols, they become the hard block. Normally, they are aromatic but when colour and clarity retention in sunlight exposure is a priority, it is best to use a hard segment aliphatic.

The soft segment can either be a polyether or polyester type, depending on the application. For example, wet environments generally require a polyether-based TPU, while for oil and hydrocarbon resistance often demands a polyester-based TPU. For greater utility, the molecular weight, ratio and chemical type of hard and soft segments can be varied. This versatility results from the unique structure of a TPU that provides high resistance, good compression, in addition to resistance to impact, abrasion, tears, weather and even hydrocarbons.

TPUs offer flexibility without the use of plasticisers, as well as a broad range of hardness and high elasticity. A main feature of a TPU is that it bridges the material gap between a rubber and a plastic. Its range of physical properties enables TPUs to be used as both a hard rubber and a soft engineering thermoplastic.

Further, TPUs can be compounded for use in numerous applications, where greater structural integrity is required, such as for automotive side body mouldings. For example, when mixed with glass fibre or mineral fillers such as mica, calcium carbonate or aluminium hydrate – here, the author would suggest that the use of bio-mass fillers like rice hull powder or bamboo flour would yield even better results for TPUs properties of abrasion resistance, high impact strength and good low temperature flexibility can all become enhanced. If rice hulls are used as ash (70–80% silica) TPU compounds can also demonstrate good fuel and oil resistance and high melt-flow characteristics, in addition to paintability. Since TPUs are outstanding contributors of polymer resistance and low temperature flexibility, when added to polycarbonate or acrylonitrile-butadiene-styrene (ABS), a TPU resin with a nominal flexural modulus of 18,000, produces compounds with flexural modulus values up to 150,000 psi. In this way, specialised compounds can be made that improves the properties of PC or ABS or other plastics as well. Some typical applications of TPUs are as follows:

- Auto-body side mouldings
- Caster wheels
- Drive belts
- Fire hose liner
- Flexible tubing
- Footwear-sports shoes
- Hydraulic hoses
- Hydraulic seals
- Inflatable rafts
- Medical tubing
- Swim fins and goggles
- Wire and cable coatings
- Coated fabrics

2.7 Filled foams

Highly filled PUR foams have their uses where foam properties are not a priority. These would be heavier than conventional flexible foams and some may even call them 'cheap foams'. A good maximum fill would be 100% of polyol or a polyol/graft polyol system used. The use of inorganic fillers such as barium sulphate or even a non-traditional compatible biomass filler is important in the production of highly filled foams. While the filler used should be compatible with the polyol, it must also completely dissolve in the polyol to make a workable 'slurry'.

If the mixing/foaming process is manual, it will not pose any problems but if a dispensing machine is used, the pumps must be able to deliver the slurry smoothly. To achieve this end, the following are recommended:

- Filler particle size to be < 5 microns
- Water content should be < 0.2%
- Filler/fillers should not be hygroscopic
- Be free of any metal parts (affects catalyst activity)

Some of the common uses for these foams are gym mats, carpet underlay and mattress bases.

2.8 Composite foams

Rapid advances in processing technology for PUs, especially in reactivity control have made PUs an ideal material for composite applications long dominated by unsaturated polyesters and vinyl esters. In the last few years, PUR composites have made in roads primarily in foamed structural RIM automotive interior and exterior parts. Such applications have gained PURs a good percentage of the long-fibre and continuous-fibre composite market. Non-foamed high-density PUR composites are also being used in combination with these products.

PUR composites are generally produced with rigid thermoset resins, as opposed to elastomeric or TPU. Composites manufactured from these PUR resins have superior tensile strength, impact resistance and abrasion resistance compared to composites based on unsaturated polyester and vinyl ester resins with the superior toughness of PUR composites paying off in secondary operations such as drilling, machining and assembly. Machined or punched edges exhibit little or no micro-cracking compared to traditional thermoset.

PUR composites are also preferred for their processing advantages. Cure times are much faster than for polyester spray-up in non-automotive applications and less labour-intensive than for polyester spray-up. In automotive parts, PU structural reaction injection moulded (SRIM) takes only 30 seconds to 2 minutes versus 2 to 10 minutes for polyester and vinyl ester sheet moulding compound (SMC). Another important aspect is the lower tooling costs because of lower operating pressure for SRIM. There has also been a downside to the reaction speed of PUs. A drawback to producing large reinforced PUR parts in the past has been the fact that the chemical reactions were so fast that it often did not allow sufficient time to close the press. However, advances in PUR technologies and new equipment designs have overcome these difficulties.

While some technologies may not allow the use of PUR resins for long open times of more than a few hours. The fast reactivity of PUR makes it a good candidate for some open-mould processes, such as spray-up of tubs, provided appropriate engineering controls are in place. PUR composites have another advantage in that they contain no styrene and do not generate large amounts of volatile organic compounds (VOCs). On the other hand PUR does contain MDI, which is a regulated material. However, tests carried out Bayer sources have reasons to believe that MDI emissions from PUR composite processing is negligible due to the low vapour pressure of MDI.

2.9 Structural foams

Advances in structural reaction injection moulding materials and processing equipment have made this the fastest growing area of PUR RIM and a prominent alternative for processors of composites using RTM, spray-up and SMC. Traditional SRIM has a lot in common with RTM in that it is a closed-mould process where a glass preform or mat is placed in a mould, which is then closed and the PUR chemicals are injected but newer advances in RTM have turned it into more of an automated spray operation.

Alternative processing methods have been long fibre injection (LFI) from Krauss–Maffei, InterWet process from Canon and composite spray moulding (CSM) Baydur from Hennecke-material science division. Each of these processing methods have also offered variants of these processes that allowed the use of natural fibres. The use of bamboo fibres instead of glass fibres or others may be more beneficial in provision of enhanced properties and cost-effectiveness. More recently, Bayer introduced three more CSM variants to the American market. Two of them are CSM-Baypreg and CSM Baypreg NF. The former produces sandwich panels consisting of a paper honeycomb combined with glass-fibre mats that are impregnated with PUR chemicals sprayed on both sides of the lay-up. The laminate is then compression moulded and cured under heat. These panels are generally accepted as greater lightweight potential than other sandwiched products, making them ideal for automotive and other applications.

The CSM–Baypreg NF (natural fibre) process is quite similar but can yield thin-walled and extremely lightweight automotive components made from natural fibres without a honeycomb core. Vehicle inner door applications are a good example.

Another of Bayer's processes is the CSM-Multitec short fibre PUR systems, a version of open-mould spray-up. A PUR mixture is applied in several layers, sold or foamed, with or without reinforcement and is allowed to cure in an open mould. In this process, glass fibres are chopped to 5–12 mm (0.2–0.5 in.) in length in an external chopper device mounted on the PUR spray gun. Traditional PUR equipment manufacturers are adding glass choppers to their two-component PUR equipment and polyester spray equipment makers are adapting their equipment for PUR also.

Although foamed low-density SRIM products have dominated PU composites for several years, particularly in auto interior applications such as door panels and others. More recent developments have shown that non-foamed, high-density SRIM systems are also viable. Examples are cargo boxes of pickup trucks, the inner mid-gate panels of utility hybrid vehicles. A non-automotive breakthrough application for large 'solid' PUR long-fibre systems can be used to form the outer skins of these doors using the Krauss–Maffei LFI system. The lower moulding pressure makes this process more economical than compression moulding.

2.10 Foams for comfort

Here, the author would like to present four types of foams – all used for primary purpose of comfort applications. First, it was latex foam and then in the 1950s, PU foams came on the market, followed by memory foam. The latter two products being petro-based, saw a resurgence of latex foam being used primarily for bedding. However, even today, the preferred material is standard PUR foam and memory foam. More recently, a new concept has seen the manufacture and emergence of a material made with natural latex, with claims of similar properties to memory foam called – *natural memory foam.*

2.10.1 Latex foam

The sap from rubber trees in the form of a liquid latex has been a source of raw material for many products for a very long time. Latex foam is produced by vulcanising liquid latex into a solid foam. The vulcanising process cross-links latex particles with sulphur through the application of heat and accelerators such as zinc oxide. Latex foam can also be produced from synthetic latex, which is produced from petrochemicals.

Latex is a stable dispersion of polymer microparticles in an aqueous medium. It is found in nature but synthetic latex can also be produced by polymerising a monomer such as styrene that has been emulsified with surfactants. The raw latex collected from the rubber trees is generally put into drums or tanks and chemicals are added and mixed to prevent coagulation.

There are many types of latex foam mattresses such as pure late foam (soft, medium and hard), latex coir mattresses, where the latex acts as a binder, hybrids and so on. Opinions may differ but as in every product, there are limitations and the end purpose will decide the quality needed, with costs also a factor.

2.10.2 Polyurethane foam (Open cell)

Standard open-cell PU foams are primarily made up of compounds derived from petrochemicals. Perhaps at the beginning, people would have had the notion that the primary uses of flexible PUR foams are for furniture but as this material became known and its uses expanded for comfort zone applications, PUR foam mattresses slowly replaced mattresses made from latex foam and spring-coils.

Many types of combinations using foam, spring-coils, latex foam, coconut fibre/latex and others are available today. Of these, the basic flexible PUR foams probably have the largest section of the market. Their two basic marketing tools are (1) density and (2) IFD (support factor). Additional properties can be additives incorporated to counter UV action, fire hazards and microbial action. Densities can range from 12 to 50 kg/cu.m, while lower densities can be produced with an additional secondary blowing agent like methylene chloride, where water would be the primary blowing agent. These materials will be very soft and have random larger voids to promote absorption and probably end up as sponges and such.

2.10.3 Memory foam

Memory foams also known as viscoelastic foams are made from special grades of polyols and are basically PU foams with special properties. There are many types and brands of these foams in the market but a good memory foam is *Tempurpaedic*. These are speciality foams much denser than the normal standard flexible PU foams with four unique properties – *density, time, temperature* and *hardness*, all contributing to ultra-comfort. If, one were to press a hand onto a memory foam surface, it would sink into the foam, leaving an impression of the hand. When the hand is removed, the foam would gradually recover its original flat surface. The softness and recovery time can be easily varied at the time of formulating to meet any particular market demands. Primary reasons for ultra-comfort when using these foams as mattresses are the automatic shaping of the foam to one's body shape because of the body heat irrespective of size or without any reverse pressure and cushioning it. These foams are naturally more expensive than standard PUR flexible foams and if one wishes, 2–3 inch thick (5–7.5 cm) mattress toppers can be used on existing bed surface, which will also give good comfort. A simple but effective way to check the quality of a memory foam is to take a small sample of it and keep it in a freezer for one hour. After this period, if the sample is rock-hard the foam is of good quality. Same test done with a standard flexible foam will show no change.

2.10.4 Natural memory foam

Memory foam mattresses made from natural latex as an alternative to synthetic memory foam, *Essentia*. The manufacturers claim that since these products are made from plant-based natural raw materials, they 'breathe' better allowing a cooler sleep than the popular mattresses made from petro-based chemicals. They also offer a 20 year life span.

These companies have developed their own technology. Basically, there are two methods of preparing latex emulsions – one developed by Dunlop which produces a 'firmer' latex and the other, a more recent innovation called – Talalay – which produces a 'light and fluffy' latex. Essentia memory foam mattresses uses the Dunlop latex method. According to them, the latex emulsion is imported from Thailand and their recipe includes small percentage of jasmine essence, cone flower oil and grapefruit seed extract and other additives as needed, poured into moulds and oven-cured. Since these products are made from natural plant-based raw materials, they are bio-degradable, which is also an advantage.

2.11 Integral skin foams

Integral skin foams consist of a two-part PU system that combines a light-weight flexible foam core encased in a thick/thin outer 'skin' that is created in one single moulding process. A complete product in a single moulding opera-tion translates into faster production times, lower labour costs and improved productivity. Depending on customer's needs, the outer skin surface finishes of these integral foams can feature decorative surface finishes, multiple colour options as well as incorporation of functional, structural or other substrates and inserts in the same single item.

Core foam densities and textures can be formulated to range from very soft to very firm, depending on the functions of the finished products. These foams can include UV stabilisers, anti-bacterial or any other via the addition of suitable additives. Depending on the end applications, these foams can also be formu-lated to meet anti-smog or environment requirements. Such a system that could be labelled as a 'green product' is – Ecoflex PUR integral skin systems made by FSI Foam Supplies Inc. Among other systems available in the market, Elastofoam two-component systems made by BASF and Dow are also widely used in industry.

Integral skin foams, also known as 'self-skinning foams' are used in a number of market applications, including automobile interiors, furniture components, household leisure goods and health care applications, such as wheelchairs. Some of the main performance benefits of these PUR integral skin foams are as follows:
- Lightweight, flexible and comfortable to the touch
- Good flow and density distribution of foams
- Safety and durability
- Superior skin properties
- Good abrasion and chemical resistance
- Excellent mechanical and dimensional stability
- A curing profile for fast de-moulding
- Easy incorporation of additives to meet most end application needs
- al skin foams are as follows:
- niture components heeds

Bibliography

1. BASF Polyurethanes: www.polyurethanes.asiapacific.basf.com
2. Sherman, Lilli Manolis – article on 'Polyurethane Composites: New Alternative to Polyester and Vinyl' – 3/1/2006
3. Rust, Cathy-article on 'Essentia Natural Foam Mattresses' – July 24th 2013: www.becgreen.ca/2013/07/essential-natural-memory-foam-mattresses
4. Mearthane Products Corporation: www.mearthane.com/products/urethane-microcellular
5. Lee, Shau-Tamg Dr. – article 'From Cellular to Microcellular' – 03 May 2004: www.plasticstrends.net
6. BASF Polyurethanes North America: www.polyurethanes.basf.us/products/name/cellasto
7. Rogers Corporation: PORON Microcellular Urethanes: www.rogerscorp.com/ems/poron
8. Bergad Speciality Foams & Composites: High Resilience Foams: www.bergad.com/us/products/highresilience-foams
9. Sheela Foam Limited: High Resilience Foams: www.pufoammanufacturers.com/high-resilience-pu-foam.html
10. FSI foam supplies Inc. – Integral Skin: www.foamsupplies.com/products/integral-skin
11. The Dow ChemicalCompany: www.dow.com/en-us/polyurethane/markets/consumer-comfort-materials/integral-skin-foam
12. SPRAYFOAM: www.whysprayfoam.org/high-pressure-two-component-spray-polyurethane

3 Two-component systems

3.1 Introduction

In this chapter, the discussions and presentations are based on the type of two-component polyurethane (PU) systems currently available, different grades, their brand names and functions for three major industrial applications highlighted as examples. Also presented are solvent-based and water-based systems in addition to more advanced systems. The author feels that this practical approach is more beneficial to readers than presenting chemical structures and analysis of the components and systems. The grades and types presented are for demonstration only and does not imply that they are the best in the market for these particular applications, although they are excellent products.

Unlike plastic raw materials in general, where common ones like polyethylene, polypropylene, polyvinyl chloride, polystyrene, polycarbonate and many others are available mostly in solid forms such as granules, pellets, powders or even as liquids, raw materials for PUs being liquids are referred to as *systems*.

A basic system would comprise at least a polyol, isocyanate, blowing agent, stabiliser, catalyst and surfactant, with an additive for colour being an option. These systems can also contain many other additives as fire retardants, anti-UV agents, anti-microbial agents or others, depending on the requirements of the end application.

There are many types of two-component systems available from traditionally reputed manufacturers like – BASF, Bayer, Dow, Huntsman and some others, while the advent of newer chemical companies are making it possible for wider range of versatile PU systems. Emerging technologies over the past few years are also making it possible for the use of biomass materials from natural sources to be used as fillers, reinforcing agents and other uses, allowing the chemical industry to be less dependent on petro-based or earth depleting materials.

3.2 The concept of two-component systems

PU technology is based on complex polymer chemistry involving several components and it is likely that any large volume producer will have a team or at least one polymer chemist in its organisation. This will allow for on-site formulating laboratory trials and perfecting many formulations as per requirements of varied customer requests.

A small volume producer or an entrepreneur with limited resources may well have to depend on the chemical suppliers, thus limiting the range of manufactures. Bulk purchases of chemicals will be more cost-effective than just-in-time (JIT) purchases which will also carry the possibility of uncertainty of supplies on

https://doi.org/10.1515/9783110643169-003

time. Manufacturing on production floors and shipping out may be a normal flow but when it comes to insulation of buildings, road applications and other on-site applications, other approaches are called for. Thus, the emergence of portable systems from very small packs to medium to bulk systems for on-site application.

The chemical industry with constant research and development has always strived to improve products, provide the markets with newer materials with better functions and properties. The PU chemical industry has seen rapid expansion and improvements and emerging technologies has been seeking portable systems and methods for on-site application. In a nutshell – the concept of two-component portable systems is based on multiple components being condensed to two basic components as – component A (isocyanate) and component B (polyol +) where simple mixing methods can be employed from hand-mixing to machine dispensing, depending on the application and applied with more or less instant curing.

Some of the common and important areas of applications are as follows:
–   Spray foams for insulation
–   Block foams for comfort products
–   Appliance foams for insulation
–   General purpose foams for various products
–   High-density foams for speciality applications
–   Surfboard foams for leisure and sports activities
–   Packaging foams for in-situ packaging
–   Taxidermy foams for speciality applications
–   Panel foams for office/commercial panelling
–   High-resilience foams for cushions/moulded seats
–   Flexible spray foams for speciality applications
–   Integral skin foams for moulded products
–   Elastomeric foams for speciality applications
–   Microcellular foams for footwear/sports goods

Advantages of two-component systems are as follows:
–   Systems are supplied as component A and component B instead of separate multiple components.
–   Suppliers take responsibility for accuracy of weighing/quality
–   Most systems available are well researched and made to acceptable standards
–   System users also have the option of custom-made systems
–   Full technical data regarding a system will be given by the supplier
–   Ease of mixing and dispensing
–   Short 'tack time' and curing times
–   Systems available in small to large packs and in some cases include spraying units as a complete package

- System users have the options of JIT purchasing, avoiding carrying large inventories
- Overall cost savings

Disadvantages of two-component systems are as follows:
- One system can produce only the formulated density
- Each system will produce only the properties designed for
- Shelf-life may be limited because of some components. For example, catalysts

Note: In some systems, it is possible to vary the density (within reasonable limits) by adjusting the proportions of component A and B.

3.3 Systems used for general applications

There are many producers and suppliers of two-component PU systems in the market with chemical giants like – Bayer, BASF, Dow, Monsanto, Huntsman and others leading the way. One can use their standard systems or request for systems as per individual requirements. Although, there are more than one brand of systems suitable for the same application, it is not possible to present most of them. Therefore, ten selected brands/systems are presented, covering most common and important applications.

3.3.1 Systems for producing flexible foam blocks

The following two systems are tried and tested ones as designed and formulated by the author and produced by a chemical company called – Premilec Inc. in Montreal. The processing methods for these two special systems will be presented in detail later in another chapter.

3.3.1.1 Maxflex 421 system
This is a fully formulated two-component system for making flexible PU foam blocks using water as the blowing agent. It is designed for production of furniture foam products, using either open- or closed-mould pour technique.

Liquid components properties:

	Component – A (isocyanate A-117)	Component B (Maxflex 421)
Viscosity at 25 °C (77°F)	210 cps	400–600 cps
Specific gravity	1.19	1.03
Mixing ratio	50	100

Reactivity profile:

Cream time (pale yellow)	12–16 seconds
Gel time	95–105 seconds
End of rise	110–120 seconds
Density-free rise	72–76.8 kg/cu.m (4.5–4.8 lbs./cu. ft.)

Typical processing conditions:

Component temperature	70–90 °F
Mould temperature range	90–125 °F
De-mould time	7–10 minutes
Iso/polyol ratio range	0.45–0.55
Moluded density range	5.2–5.6 lbs./cu.ft

3.3.2 System for high-resilience foams

Grade – Greenlink HR 250 is manufactured by ERA Polymers Ltd. (Australia). This two-component water-blown system is designed for making high-resilience PU foam for cushions for office and domestic furniture. This system can be manually drill-mixed at a minimum speed of 3,000 rpm but for large volume productions, it is preferable to process through a dispensing machine.

Component properties:

	Polyol	Isocyanate
Appearance	Opaque liquid	Brown coloured liquid
Brookfield viscosity (cps)	1,450	450
Specific gravity	1.05	1.20
Isocyanate value NCO (%)	–	25.5

Reaction profile:
Based on laboratory results on hand-mix @ 20 °C
Mix ratio by weight (Polyol: Iso) = 100:57

Mixing time	7–8 seconds
Cream time	12 seconds

Gel time	70 seconds
Tack-free time	300 seconds
Free-rise density	57 kg/cu.m

Note: The ratio of Greenlink HR 250 can be changed to produce harder or softer foam.

3.3.3 Spray foam systems

Foamsulate brand two-component PU spray systems manufactured by Accella Polyurethane Systems LLC (USA) are energy-efficient spray foams for insulation of residential, commercial and industrial applications.

Spray polyurethane foam (SPF) insulation is one of the most innovative and versatile developments in building construction and insulation technology in the past few decades, quickly becoming the insulation material of choice for builders, architects and homeowners. Unlike fibreglass, cellulose or other loose-fill insulation products, Foamsulate systems can be applied with precision, on-site anywhere, expanding to completely fill gaps and spaces with a long-lasting solid insulation. No matter, how irregular the shape, how rough the surface or how difficult access is, this insulation will flow quickly filling all areas and create a perfect fit.

Some of the common applications are follows:
- Exterior walls
- Vented attics
- Between floors
- Uninsulated basement walls
- Foundations
- Crawlspaces
- HVAC ducts
- Fluid tanks
- Cold storage units

According to Accella, they have three major types of spray foam systems.
1. **Light-density spray foam (open-cell)** – is a cost-effective solution that delivers a high R-value per inch with low air filtration properties for a thermal envelope that is much more tightly sealed than traditional fibreglass, cellulose or other loose-fill products.
   - Low air filtration reduces pollutants, dust and allergens from outside
   - Open-cell structure delivers noise reducing qualities
   - Cost-effective, builder-friendly option
   - Commonly used in exterior walls, between floors and so on

2. **Medium-density spray foam (closed-cell)** – because medium-density spray foam has a higher R-value per inch, it will act as a complete air and vapour barrier for a thermal envelope that is lighter-density foams and far superior to traditional forms of insulation.
   - Completely blocks pollutants, dust and allergens
   - Cures to a semi-rigid solid that adds significant strength to structures
   - Commonly used in exterior walls and vented assemblies between floors, foundations, crawl spaces, fluid tanks and cold storage units

3. **High-density spray foam (close-cell)** – these foams work exceptionally well in commercial environments, as they are very cost-effective for use in large spaces. Superior insulation properties and high rigidity make it the ideal spray foam product for areas that require a tougher, more durable application.
   - Allows for advanced temperature control
   - Improves structural strength
   - Frequently used in the agricultural industry

3.3.4 Systems for integral skin foam moulding

Dow Polyurethane has announced a new PU system for integral skin foam moulding applications with improved mechanical and physical properties starting late 2016. This self-skinning grade will help lower global warming potential (GWP), while still sustaining the comfort levels of previous PU systems.
   Performance benefits:
- Lightweight flexible foam core but soft to the touch exterior
- Good flow and uniform distribution of foam density
- Safety and durability
- Superior skinning properties
- Good abrasion and chemical resistance
- Excellent adhesion to a variety of substrates
- Excellent mechanical properties of the foam resulting in high-dimensional stability
- Extended shelf-life of fully formulated system with hydrofluoroolefin (HFO)
- A curing profile that allows fast de-moulding times
- Additives can be added – flame retardant, UV stability, anti-microbial needs and colour

3.3.5 System for automobile applications

Elastoflex E manufactured by BASF is a tailor-made semi-rigid PU foaming system used for foam backing films, skins or leather in automobile applications. The economic benefits can be primarily attributed to its short de-moulding times, material and energy savings and high process safety factors.

Special features:
- Improved passenger protection against impact
- Provide hidden airbag solutions
- Single shot foam backing for a highly complex moulding geometry
- Good adhesion to films, skins and instrument panel beams
- Weight saving obtained by low-density semi-rigid foam
- Short de-moulding times
- Excellent elasticity and good thermal and anti-freezing resistance of finished parts
- Defined soft touch for finished parts
- Provides additional noise and vibration absorption
- Freedom of design for more distinct contours

3.3.6 Systems for insulation of pipe joints

This is a speciality application. PU systems used for pipe insulation for heating must fulfill various stringent requirements concerning the end properties and processing performance. A low foam thermal conductivity and a good pipe composite heat resistance are important, as network heating services temperature can go up to 130 °C (266 °F) with occasional peaks up to 140–150 °C (284–302 °F). The foam system must have good flow properties which enables the foam to fill narrow cavities. Excellent adhesion is also required to ensure efficient long-term performance.

Huntsman Polyurethanes is a major supplier of PU systems for these applications. They have developed a series of fully water-blown systems for insulation of joints and also pre-insulated pipes. These systems consists of two-component liquids that foam to give excellent insulation with highly comparable strengths, durability and efficient long-term service. The quality of the foam depends greatly on the right mixing ratio. To avoid errors in polyol and isocyanate weighing on-site, the pre-prepared two-component systems are a great advantage and the use of on-site foaming machines will ensure accuracy and material savings.

Daltofoam TE 44214 and Daltofoam TE 44205 are fully formulated two-component polyol blends consisting of polyether polyols, silicone surfactant, catalyst and water. These two grades are specially designed for the production of pipe joint systems and will give excellent ageing characteristics. System TE 44214 is particularly designed for pipe joint insulation for very low temperatures.

Product characteristics:

Property	Units	Daltofoam TE 44205	Daltofoam TE 44214
Appearance		Pale yellow liquid	Pale yellow liquid
Hydroxyl value	Mg KOH/g	575	550
Viscosity @ 25 °C	mPa.s	930	450
Density @ 25 °C	kg/L	1.08	1.08
Flash point	C	>100	>100
Mixing ratio: polyol/MDI	Parts by weight	100:193	100:188
Reactivity/cream time	Seconds	71	50
Free-rise density	kg/cu.m	51	36

3.3.7 Systems for structural and casting foams

BASF polyurethane systems include many products and *Elastolit* is a very versatile rigid foam two-component system for producing structural foams. Basic specifications are as shown below:

Product group	Polyurethane structural rigid foam
Product name	Elastolit
Applications	Technical parts Decorative parts Automotive parts Wood imitations
System characteristics	Excellent flow Good hardness at low density Good de-moulding properties High scratch resistance
Fire performance	According to DIN 4102 82 and 83
Blowing agent	Carbon dioxide
Density (kg/cu.m)	100–1,100

Basic specifications for casting foams are as shown below:

Product group	Non-cellular polyurethane casting compound
Product name	Elastocoat C6

Applications	Decorative parts
	Edging processing
	Rotation moulding
	Plates (glass fibre reinforced)
System characteristics	Good flow – high impact filler
Filler (if applicable)	Inorganic
Hardness	System dependent – controlled by mixing ratio (within small limits)
Moulds	Systems are suitable for all kinds of moulds – Silicone, epoxy, aluminium

3.3.8 Systems for thermoplastic foams

BASF offers a new type – *Elastollan* AC 55D10 HPM (high-performance material), a TPU that combines freedom of design and long-term durability with a high-class appearance. This new grade not only boasts the properties of conventional TPUs but shows new features as well. BASF has also further developed the processing properties of this new Elastollan grade. This material can be injection moulded within a wide process window with a cycle time 50% shorter compared to conventional TPU. It has excellent flow properties, which means that wall thicknesses of 1.5 mm to 1.8 mm can be produced without affecting the surface quality.

Key properties:
- Good tensile strength and abrasion resistance
- Excellent impact resistance at low temperatures
- Good elasticity and scratch resistance
- Outstanding surfaces even at thin wall thicknesses
- UV and weather resistance
- Easy processing and easy cleaning
- Good colourability
- High service temperatures– 120 –150 °C

One of BASF's innovations is a grade of **super soft plastics – TPU without plasticiser** – with exceptional texture and 'good feel'. The handles of electric tools, cellular phones, laptops and others have a common requirement, namely a nice dry feel with a good texture. This material provides them and much more. The secret lies in the formulations of the materials based on an innovative range of soft plasticiser-free thermoplastic PUs, which fulfills these characteristics.

Material combinations of two different hardness are often used in order to combine different properties of plastics. Thermoplastic elastomers (TPE) are used in many cases as soft components. Generally, it has been necessary for the production of thermoplastic PUs with hardness levels below 75 Shore A to use plasticisers. Migration of the plasticiser can, however, lead to increase in the hardness of the final product. In automobiles, migration of plasticiser can cause fogging inside and finally losing flexibility. A good example is the migration of plasticiser from artificial leather upholstery and moulded parts, causing fogging.

To counter these factors, BASF has succeeded in the development of thermoplastic PU that provides low hardness levels without the addition of plasticisers. According to reports, the main basis for success is a targeted modelling of the product's molecular structures. This has enabled economical processing, good mechanical properties and softness of the final products.

3.3.9 Systems for protective coatings

Protective coatings are a must for most areas, especially in industrial applications. Some methods may take the form of dip-coating, calendaring, hand-painting but for industrial purposes, spray coating delivers a more advanced, efficient and easy-to-handle process. Two special foaming systems – *Premcote* and *Premiseal 280* two-component PUR systems made by Accella Polyurethane Systems LLC (USA) are ideal protective coating systems. Their high elasticity, resistance to abrasions, corrosion protection, chemical resistance and waterproofing properties make these coatings ideal for protection, decoration and structural enhancement.

These versatile protective coating systems are used in several industries, with applications being discovered all time. Some of the main applications are as follows:
- Roof coatings
- Storage tanks
- Secondary containment systems
- Protective coatings for bridges
- Floor areas

Special features
- Fast setting
- Easy spray application even in low temperatures
- Accurately proportioned and mixed for consistent quality
- Excellent adhesion to a wide variety of substrates – concrete, metals, wood and others UV – stable colours
- Slip-resistant additives and different surface textures available

3.3.10 Systems for memory foam (viscoelastic)

Many grades and types of memory foams have been developed over the past few years and now, a wide variety of materials are available in the market. BASF CosyPUR is the latest PU flexible foam which has unique qualities. With its specifically adjustable product properties possible by its cell morphology, CosyPUR is modifiable from ultra-soft to viscoelastic and extends the application possibilities of flexible foams. This versatile flexible foam is suitable for the production of mattresses, pillows, cushions, back-rests, orthopaedic products and many others.

Three grades of CosyPUR are available as two-component systems as shown below:
- CosyPUR Visco systems can be used to produce both viscoelastic moulded and slabstock foams
- CosyPUR Super-soft – for textures like super-soft – latex-like material for both moulded and slabstock foams
- Cosy PUR Balance – a viscoelastic super-soft system based on renewable raw materials

Polyol		Isocyanate	Others
Slab VE foam	Polyol for VE	ISO for VE	Catalyst and water
Slab HR foam	Polyol for HR	ISO for HR	Catalyst and water
Moulded VE foam	Polyol for VE	ISO for MF	–
Bio VE foam	Bio-Polyol for VE	ISO for Bio-VE	–
Super-soft (latex-like) block foam	Polyol for super-soft	ISO for super-soft	Catalyst and water

Advantages of CosyPUR Visco system:
- Pure MDI (isocyanate) system
- Easy superior processing
- Excellent physical properties even at low temperatures
- No expensive cell openers required
- Superior – very soft feel
- Wide density range available
- Low odour
- Adjustable properties

CosyPUR Super-soft:
A new super-soft, flexible foam for sports, leisure activities and total relaxation.
- Moulded density from 45 to 250 kg/cu.m
- Extraordinary elasticity

- Exceptional permeability and breathability
- Durable
- Low emissions
- Great soft texture (similar to latex)
- Odourless

CosyPUR balance:
A viscoelastic and super-soft system based on renewable raw materials. This system is an environment-friendly solution for flexible slabstock foam made with Lupranol Balance, a natural oil polyol (NOP) from the castor oil plant and can be used as a 100% drop-in for any other conventional polyol. These foams are ideal for mattresses and makes sleep even better than most foams.

Advantages of foams made with Lupranol Balance are as follows:
- Good processing profile
- Good mechanical properties
- Low emissions and odour
- 25% of renewable raw material in the foam
- Reduces the use of petro-based chemicals

3.4 Solvent-based two-component polyurethane systems

There are many grades of solvent-based PU systems currently being used and also being constantly developed. The following presentation highlights one grade for each important application for demonstration purposes only and does not imply it as the best in the market. The three areas are as follows: (1) adhesion, (2) coating and (3) coloured floor applications.

3.4.1 Solutions for superior adhesion

Desmocoll hydroxyl-terminated high molecular weight PUs are specifically suitable for formulating solvent-borne PU and polyester systems. Available as small granules solubilised in organic solvents, Desmocoll yields an end product that exhibits superior adhesion to a large array of substrates, including PVC.

Based on polyester polyols and aromatic isocyanates, Desmocoll grades differ in their crytallisation rates, heat resistance, adhesion to specific substrates and solution viscosity. Because of the wide product range offered by Covestro, it is possible to select a suitable grade for a variety of adhesive applications. Adhesives based on Desmocoll are generally used for heat activating bonding processes with bonding strengths increasing as the adhesive layer cools. Desmocoll produces maximum performance when used with Desmodur R polyisocyanates.

Desmocoll grades 400, 500 and 600 series grades enable exceptional adhesion to PVC substrates and has high resistance to the most commonly used plasticisers. They are ideal for footwear, rubber, packaging, furniture, textiles, wood, plastics, aluminium and plastic films.

3.4.2 Solvent-based coatings

NORMAC PU coatings provide excellent protection with corrosion control and abrasion resistance. They can be diluted down to enable spraying through small electric airless cup guns or gravity-fed air guns or also left undiluted for fast high-build applications with large airless spray pumps. All NORMAC sprayable systems shown below are non-flammable.

NR – 2S
Two-component, TDI ether-based protective coatings with a Shore Hardness of 50A, designed to provide a higher coefficient of friction, high elongation and high resilience. Typical coatings would be for conveyor belts and foam coatings.

BR – 3S
Two-component aliphatic ether-based PU coating systems with a Shore Hardness of 80A designed to provide high abrasion resistance. This product has over 25 years of proven application history in all types of industry, especially mining.

NR – 5S
Two-component aliphatic ether-based coating systems with a Shore Hardness of 90A designed for applications requiring load release and good abrasion resistance. Some typical applications include salt and sand spreaders, street sweeper bodies and so on.

NR – 7S
Two-component aliphatic ether-based coating systems with a Shore Hardness of 98A. They are designed specifically for load release of sticky materials. Typical applications include vacuum truck bodies, snow plow blades, boats and tankers among others.

NR – T10 Additive
A single-component additive available for all two-component urethane systems that acts as a thixotropic agent, enabling much higher build per coat and more economical application costs.

3.4.3 Solvent-based high-gloss surface coatings

Commercial and industrial floors need durable and protective surface coats, especially where high traffic occurs. Surecrete's high-gloss PU systems are high-quality floor coatings especially suited for concrete surfaces. Made and marketed under their brand name *Dura-Kote*, formulated with high-quality raw materials and UV stable pigments, these are prime choices of builders and contractors.

Dura-Kote solvent-based gloss PU floor coating is a two-component acrylic aliphatic PU. This grade is designed as a very thin colour coating for interior concrete, cement-based overlays or as top coats for epoxy systems. This high-performance pigmented coloured top coats generate strength, added UV stability, flexibility, chemical and scratch resistance in addition to being user-friendly and durable. These solvent-based PU systems are especially suitable for both commercial and residential settings, aircraft hangars and clean room floors, manufacturing facilities, warehouses, retail stores, automotive showrooms, garage floors, stadiums or any high traffic areas where an exceedingly resilient and durable floor is desired.

Selecting a correct grade among the many, especially for interior applications can sometimes be a tricky proposition, when a high-performance surface with easy cleaning and durability is required. Dura-Kote Pigmented Polyurethane SB is recommended by the manufacturer ideally suited for these applications with combinations of a durable surface and a solid coloured coating. Dura-Kote systems can also be made according to custom orders with a selection of a preferred colour chosen from the wide range of colours available.

Specifications of Dura-Kote Coloured Polyurethane Solvent-Based Floor Coating:
- Coating: 300–400 square feet per gallon
- VOC rating: less than 500 g/L
- Solid contents: 51%
- Shelf-life: 1 year (unopened container)
- Ready for recoat: 6–8 hours (18+ hours at low temperatures)
- Full cure time: 5–7 days (up to14 days at low temperatures)
- Coating appearance (cured): coloured gloss sheen
- Mechanical stability: excellent
- Light stability: excellent
- Diluent: hydrocarbons
- Odour: solvent
- Application temperature: 50–90 degrees Fahrenheit

3.5 Waterborne two-component polyurethane systems

The need for water-based coatings instead of solvent-based coatings is primarily driven by stricter regulations on volatile organic compounds (VOCs), which has been moving towards 50–100 g/L and also the desire to be less dependent on petro-based chemicals. In addition, consumer demands for low odour coatings, especially those used in interior applications. The challenge has been to make a low VOC, low-odour system that offers the performance properties of a two-component solvent-based PU coating.

Two-component waterborne PU coatings have been under development for decades and now some are available which meets the requirements for low-odour and low VOC. The quality of these systems has steadily improved to the point where their performance properties are acceptable by industrial stds.

Anabond Polyurethanes manufactured by Anabond Limited of India are versatile two-component moisture curing systems and have a wide range of grades for different applications. Some of them are as follows:

Anabond 720 is a two-component room temperature curing glossy PU floor coating for railway coaches or similar applications. It has a hardness of Shore A 85–90.

Anabond 7900 is a two-component PU fire retardant transformer potting compound. It has hardness of Shore D 75–85.

Anabond 797 is a two-component PU sealant for low voltage cable jointing. It has very good hot water resistance and very low water absorption.

Anabond 706, 707 and 708 are two-component room temperature curing transparent PU systems suitable for electronic printed circuit board encapsulation.

Anabond 794 and 7930 are two-component room temperature curing PU adhesives suitable for automobile and industrial filter bonding applications.

With growing environmental concerns and the plastic industry seeking ways to phase out the use of petro-based chemicals, the newly developing environment-friendly waterborne two-component PUs represent a new technology, combining the good properties of solvent-borne two-component PU systems with low VOCs of waterborne two-component PU systems. These waterborne systems have now developed to such an extent that they offer realistic alternatives to the popular solvent-borne ones.

Bayer AG Coatings, for one, has been developing and introducing easy to apply, rapid drying coatings with well-balanced range of properties at very low VOC level. Also included are high-performance PU clear-coats which combine good chemical resistance and other important aspects. Waterborne soft-feel lacquers are already well-established in the market.

In developing these new generation PU systems as well as application technology, some of the barriers that have been overcome are as follows: understanding the fundamental principles of aqueous two-component PU systems, like side reactions during pot life and film formation or the cause of blistering and also development of water dispersible polyisocyanates.

PU-based systems have an established place in the market, especially for coating industry; in some application areas they dominate. There are two main reasons for this. The first is that PU coatings give a very high level of quality with aesthetic values. They combine outstanding resistance to solvents and chemicals, with good weather stability. It is possible to formulate both clear coats and because of their good pigment wetting properties, pigmented topcoats give high-gloss, high-bodied films with excellent flow properties. These films have outstanding mechanical properties and also provide the ideal balance with PU coatings.

3.6 Some additional polyurethane systems for speciality applications

The following two-component PU systems are for making speciality foams for – *Integral skin foam, taxidermy foam, prototyping foam, block foam, moulding foam* and *packaging foam.* A few more PUR systems are also presented covering a wide spectrum of applications. Most of the manufacturers of these systems will also offer custom-grades, although their wide ranges of grades/systems available should cover most end applications.

3.6.1 Integral skin foam two-component polyurethane system

Ecofoam ISF 136 is an Integral Skin Polyurethane foam made by ERA Polymers (Australia) with a free-rise density of 135 kg/cu.m. This product does not contain any chlorofluorocarbons (CFCs) or HCFCs and is environmentally friendly that has no ozone depleting potential. This two-component PU system is suitable for moulding articles, where skinned foam is required and also can be pigmented. This system can be manually drill-mixed at a minimum speed of 2,000 rpm but it is preferable to process through a PU dispensing machine, especially for large parts. Figure 3.1 shows a moulded part of an integral skin product.

Figure 3.1: Moulded part of an integral skin product.
Source: Reproduced with permission from Era Polymers Ltd.

Component properties

Component	Polyol	Isocyanate
Viscosity (cps) @ 20 °C	725	145
Appearance	Hazy white liquid	Brown liquid
Specific gravity @ 20 °C	1.07	1.2

Note: Colour of polyol will change if pigmented.

Reaction profile

Mix ratio by weight Polyol:Iso	100:37
Mix time (seconds)	10
Cream time (seconds)	24
Gel time (seconds)	49
Tack-free time (seconds)	75
Free-rise density (k/cu.m)	1.35

Actual processing methods of these grades will be discussed and presented in detail in a later chapter under-Processing Methods.

3.6.2 Two-component polyurethane taxidermy moulding foam

Erathane TX 56 made by ERA Polymers (Australia) is a rigid PU foam specially designed for taxidermy moulding applications. It is a two-component PU system comprising of a PU and an isocyanate. When mixed at the correct ratio, will produce foam with a free-rise density of 56 kg/cu.m.

Component properties

Component	Polyol	Isocyanate
Appearance	Clear, honey-coloured liquid	Brown liquid
Viscosity (cps)	300	250
Specific gravity	1.15	1.22

Reaction profile

Mixing ratio Polyol:Iso	100:100
Mix time (seconds)	20
Cream time (seconds)	120
Gel time (seconds)	270
Tack-free time (seconds)	420
Free-rise density (kg/cu.m)	56

3.6.3 Two-component polyurethane systems for rapid prototyping

RenPIM 5212 A/ 5212 B from Huntsman is a fast curing two-component PU system for making prototypes. The information provided is taken from their 2004 Publication no. T133f GB. Special features include easy pigmentation, low flexural modulus, high impact strength, excellent surface finish and simulates a thermoplastic finish. RenPIM 5212 A/B is a neutral pigmentable system which simulates the end properties of high-density polyethylene. This system is ideal for rapid prototyping and short production runs. It can be used to produce functional prototype parts suitable for use in all major industrial areas such as automotive, aerospace, consumer goods, electronic and leisure applications.

Product data

Property	Unit	RenPIM 5212 A	RenPIM 5212 B
Appearance colour	Visual	Liquid amber	Liquid white
Viscosity at 25 °C	m Pa s	ca. 25–35	ca. 1,500–2,000
Density	g/cu.cm	ca. 1.2	ca. 1.1

Processing data

Mix ratio	Parts by weight	Parts by volume
RenPIM 5212 A	60	50
RenPIM 5212 B	100	100

'Parts in Minutes' PUs are specifically formulated to be fast curing systems and therefore require processing via dispensing machines. Huntsman Advanced Materials Division can advise on all types of mixing and dispensing machines suitable for processing this system. Thorough stirring to ensure uniform dispersion of materials is critical prior to processing. Hand mixing or manual processing of these materials are not recommended by the manufacturer.

These systems are compatible with most major silicone supplier's products for moulds. Huntsman Advanced Materials Division also offer alternate PU and epoxy materials for the production of moulds with appropriate recommendations for release agents and for colouring.

AG Kobiboden/8840 Einsiedeln, Switzerland. Other PU compatible pigments or dyes are also acceptable for colouring PUR prototyping systems. Generally, they are available in 50 g or 1 kg packs.

Properties

Resin/hardner mix		Unit	RenPIM 5212 A/B
Gelation time		s	ca. 100–130
Max. layer thickness		Mm	4
De-moulding time (depending on layer thickness)		mins	ca. 15–20

After cure: 14 hours at 80 °C or 7 days at 20–25 °C

Density	ISO 1183	g/cu.cm	ca. 1.2
Hardness	ISO 868	Shore D	55–65
Deflectiontemperature	ISO 75	C	45
Tg	(DSC)	C	82
Impact strength		kJ/sq.m	>70
CompressiveStrength	ISO 694	MPa	18–30
Tensile strength	ISO 527	MPa	15–25
Elongation at break	ISO 527	%	30–40
Flexural strength	ISO 178	MPa	>20
Flexural modulus	ISO 178	MPa	550–750
Linear shrinkage		mm/m	1

For further detailed information regarding storage of materials, processing equipment, maintenance and packaging, it is recommended that the Huntsman Advanced Materials Division be contacted.

Another excellent alternative for pattern makers, prototype makers and model makers is Duramould EM73D FASTCAST offered by Dow Hyperlast which is a cold castable system without the need for ovens or heated moulds and is applied in an easy 1:1 ratio. This material helps to increase both speed and quality for rapid proto-typing, pattern-making and tooling applications. With low viscosity enabling easy casting of intricate shapes, FASTCAST accepts fillers easily and controls shrinkage to less than 0.5%, while providing excellent accurate reproduction.

With a gel time of 3 minutes and a 10–15 minute de-mould time, FASTCAST also helps to increase productivity to produce natural or coloured moulds, which are easily machined. Duramould EM 73D is capable of casting complex mouldings that incorporate thin and thick sections. The moulded urethane is hard wearing and durable, offering the user a cost-effective tooling system.

Another interesting application is the sprayable HYPERKOTE EMH85A offered by Dow Hyperlast which can be sprayed onto a natural pattern to create a mould in minutes, enabling extra fast product development, especially for concrete products. This grade is applied on a 1:1 volume ratio to produce excellent detailing and finger-print reproduction. It has low viscosity and is solvent-free as well as being tough, durable and lightweight. HYPERKOTE EMH85A offers low-cost product develop-ment for precast concrete patterns.

3.6.4 Hand-pourable polyurethane viscoelastic flexible foam

Grade VEF-99 offered by Northstar Polymers using new technology has been formu-lated to create a very soft flexible foam for a variety of cushion applications. The cured foam has a very low deflection even into a deeper depression rate and it con-forms to the object with excellent viscoelastic behaviour. VEF-99 system is based on a polyether polyol PU, which stays soft even at very cold temperatures.

The component materials are low viscosity liquids at room temperature. The pot-life is above 30 seconds, which allows small quantities to be batched manually without a dispensing machine. The mixing ratio is set for 1:1 by volume for easy metering. This makes VEF-99 a more user-friendly viscoelastic material for product designers, prototype makers and short-run or small-volume producers. These prop-erties of VEF-99 will enable product designers to create unique foam parts and shapes to be easily moulded to make cushioning parts and make many speciality products.

The processing of this new technology material will be presented in full detail under the Processing Methods chapter.

3.6.5 Microcellular polyurethane elastomers for auto components

AUTOTHANE is an exceptional microcellular elastomer material specially designed and developed by Dow Hyperlast for the manufacture of automotive suspension components. The production of AUTOTHANE automotive components is done by Harita Seating Systems Limited (HSSL) of Tamil Nadu under licence from Dow. The inherent dynamic performance of PU enables the closed-cell structure of AUTO-THANE to demonstrate enhanced performance versus traditional materials through its ability to withstand repeated high level compression, while retaining its original state. As these components are compressed, they suppress and absorb shock and vibration to help improve the vehicle's ride profile. Their resilience enables them to repeat high performance for a long time.

In addition to excellent hydrolytic stability and dampening properties, AUTO-THANE also demonstrates exceptional resistance to chemicals, grease, ozone and microbial attacks, as well as to water, salt, oils and fuels. It is also virtually unaffected by extremes of temperature or climate. When combined with innovative mould design and efficient manufacturing processes, AUTOTHANE material is an ideal choice for mass production and is used in many applications to help isolate road noise and vibrations such as jounce bumpers, spring aids, seals, gaiters and tie rod isolators.

3.6.6 Protective coating systems for the mining sector

Dow Hyperlast offers two new materials – (1) DIPRANE 58 and (2) HYPERLAST 110 aimed at helping mining and materials handling component manufacturers comply with the European Restriction of Hazardous Substance standards. Both polyester-based DIPRANE 58 and polyether-based HYPERLAST 110 provide excellent mechanical properties. The former is a three-component quasi-PU system with a hardness range from 45 to 90 Shore A. Its main application is in material handling such as mineral and ore handling and also offers high modulus, excellent tear and abrasion resistance across the hardness range. This system is also available as a two-component system.

HYPERLAST 110 is a three-component quasi-PU system with a hardness range from 60 to 90 Shore A. In finished moulding applications, this material offers excellent mechanical, abrasion and tear performance, while its polyether composition delivers excellent hydrolysis resistance making it especially suitable for use in humid and damp conditions. This grade is also available as a two-component system. Customers can easily pigment bot systems, although Dow Hyperlast will provide customised colour versions for high volume sales.

3.6.7 Two-component polyurethane system for tabletop protection

A speciality protective coating for rail travel comfort and similar applications – DURELAST is available with Dow. These systems will produce aesthetically pleasing products like tabletops and others by providing durable PU edging for carriage tabletops and composite panels. Moulded onto a tabletop or panel, DURELAST creates a very strong bond and a hermetic and hygienic seal that is not easily found in other materials.

The combination of durability with design freedom is unique providing tremendous opportunities for any designer of cabin, carriage interiors and so on. A two-component PU elastomer available in many colours, DURELAST can be moulded to any shape, an excellent choice for smooth surfaces requiring protection against wear. It also provides a waterproof seal, easy to clean, non-toxic, resistant to heat and chemicals and helps keep surfaces looking good for long periods of time.

3.6.8 Polyurethane system for marine applications

Visibility is paramount for the effectiveness of marine navigation buoys and PU grade HYPERLAST 100 is the prime choice of Herikon BV, design specialists, developers and production of customised, high-grade technical PU products. One of their speciality manufactures are marine buoys, and they use this grade to ensure that the day marks on the navigation buoys it manufactures remain visible, virtually at all times.

HYPERLAST 100 is tough, wear resistant and provides excellent impact strength. It is UV stable and colour-fast for many years – two properties essential for navigation buoys. The day marks produced with HYPERLAST 100 comprise upper visible element of a navigation buoy. This grade of PU is an alternative to steel, not only helping to reduce costs but also makes them easier to handle because of the reduction in weight.

Unlike steel, HYPERLAST 100 produces better performance in low temperatures with virtually no ice able to adhere to the PU surface. This gives high visual identification even in poor weather conditions and maintains good buoyancy in icy temperatures. HYPERLAST 100 allows efficient production with its fast de-moulding time with moulds running below 95 °C. The surface of the urethane will accept both paint identification and vinyl lettering. Day marks are produced either in green or red but Herikon has developed an optional two-colour combination of yellow and black which could be moulded. Herikon is a world leader in marine applications in the Netherlands.

3.6.9 Two-component polyurethane system for block foam

Erathane BS22 from ERA Polymers Ltd. (Australia) is for making rigid PU block foams for board stock and pour in place applications. This formulation contains fire retardants and has a free-rise density of 33 kg/cu.m. The degree of insulation is determined by the thickness of the foam used. For cavity fill or moulding applications, it is recommended to mould to a density of 36–38 kg/cu.m. At temperatures less than 15 °C (59 °F), the reaction rate of Erathane B22 will be much slower resulting in an increase in density, reduction in foam yield and quality. Under these conditions, the use of drum heaters or temperature controlled conditions is recommended for drum storage at temperatures 18–25 °C (64–77 °F).

Erathane B22 can be manually drill-mixed at a minimum speed of 2,000 rpm or processed through a low-pressure foam dispensing machine. Machines from GUSMER and CANNON are recommended. This foam is suitable for a wide range of insulation, buoyancy or cavity filling.

Component properties

	Polyol	Isocyanate
Appearance	Clear, light amber liquid	Clear, brown liquid
Viscosity (cps)	700	250
Specific gravity	1.125	1.22

Reaction profile

Mix ratio by weight (polyol:iso) 100:100

Mix time (secs.)	20
Cream time (s)	45
Gel time (s)	225
Tack-free time (s)	375
Density (kg/cu.m)	33

Always wear safety equipment according to the Material Safety Data Sheets (MSDS) for two-component PU systems. Main protections should be for eyes and skin contact.

3.6.10 Two-component systems polyurethane systems for rigid moulding foam

Erathane MF95 is a specially formulated polyol and isocyanate system with a free-rise density of 95 kg/cu.m for rigid mouldings. This system is designed for use in moulding applications where good skin is required. This foam can be used for moulding components such as imitation wood, computer cabinets, shoe heels and fishing lures. At temperatures less than 15 °C (59 °F) the reaction rate of this system will be much slower resulting in an increase in density and reduction in foam yield and quality. Also, at temperatures above 30 °C (86 °F), the cream time will be reduced drastically.

Component properties

	Polyol	Isocyanate
Appearance	Clear, honey coloured liquid	Dark brown liquid
Viscosity (cps)	580	250
Specific gravity	1.05	1.22

Reaction profile

Mix ratio by weight (Polyol: Iso) 100:105

Mix time (secs.)	20
Cream time (s)	50
Gel time (s)	165
Tack-free time (s)	195
Free-rise density (kg/cu.m)	95

These components are sensitive to humidity and at all times must be stored in sealed drums/packs. At elevated temperatures problems may arise with pressure build-up inside the drums. When opening drums, extreme care must be exercised in releasing the internal pressure inside and also it is recommended that the contents inside should be mixed well before use.

3.6.11 Two-component systems for high-yield packaging foam

Greenlink EraPak from ERA Polymers is a two-component high-yield packaging PU foam system with a free-rise density of 9 kg/cu.m. This system has been specifically designed for packaging where the cushioning properties protect products during transit, warehousing and handling. EraPak, an alternative packaging is economical and efficient, expands and forms the shape of a product for ultimate protection. This system can be applied through any standard foam dispensing equipment.

Component properties

	Polyol	Isocyanate
Appearance	Opaque liquid	Brown liquid
Viscosity (cps)	1270	200
Specific gravity	1.003	1.23

Reaction profile

Mix ratio by weight (polyol: Iso) 100:120

Mix time (secs.)	8
Cream time (s)	10
Gel time (s)	40
Tack free time (s)	55
Free-rise density (kg/cu.m)	9

At temperatures less than 15 °C (59 °F), the reaction rate of Green EraPak will be much slower, resulting in an increase in density, reduction in foam volume and quality. At temperatures above 30 °C (86 °F), the cream time will be drastically reduced. Safety precautions must be taken as per MSDS data sheets.

Bibliography

1. ERA Polymers Pty. Ltd. – Two-Component Systems: www.erapol.com.au
2. Premium Spray Products – Polyurethane Spray Foam Roofing Systems: www.premiumspray.com/systems/roofing
3. Normac Adhesive Products Inc. – Solvent Based Elastomeric Coatings: www.normacadhesives.com
4. COVESTRO – Solventborne Polyester Polyurethane Systems: www.adhesives.covestro.com
5. SURECRETE – Solvent Based High Gloss Polyurethane Systems: www.surecrete design.com
6. Peter Schmitt, Kathy Allen and Joe Pierce – Article on Two-Component Waterborne Polyurethane Coatings, the Waterborne Symposium, Feb. 24–28 2014, New Orleans, LA.
7. Anabond Limited-Two-Component Moisture Curing Polyurethane Systems: www.anabond.in/polyurethane-adhesive.html
8. Northstar Polymers – Hand-Pourable Viscoelastic Flexible Foam: www.northstarpolymers.com
9. BASF-CosyPUR Systems: www.polyurethanes.asiapacific.basf.com
10. Huntsman – RenShape Solutions – Two-Component Polyurethane System for Rapid Prototyping: www.renshape.com
11. Dow Hyperlast – ADVANCE An Update in Polymer Engineering: www.hyperlast.com

4 Basic raw materials for polyurethanes

4.1 Raw materials

Polyurethanes (PUs) are one of the most versatile and important polymers of plastics and belong to the *thermosetting group* meaning that it cannot be re-used. However, it can be recycled and also brought into ongoing production within limits. PUs are produced based on complex polymer chemistry, with several chemicals combining to form a stable urethane. When PUs came on the market, its most popular uses were for comfort applications but because of its versatility and unique properties it quickly spread to most consumers, industrial, automotives, engineering and many other applications, including space travel, over a short period of time.

Although the raw materials are of chemical nature, it is an advantage for the reader to have a basic knowledge of them. The author will present these materials in simple terms in addition to their functions. Since these materials are classified under 'hazardous materials', information is provided for proper handling, storage and safety factors to be observed. Even though this presentation is mainly about ready-made two-component systems, which lessens the hazardous factor for a product manufacturer, it is essential to be aware of safe handling and processing procedures.

The information provided of non-traditional fillers based on emerging technology should also be of interest. PUs are formed by a basic polyol and an isocyanate with either chemical or physical blowing to make foams. Other additives are water (blowing agent), fillers, extenders, activators or positive catalysts to hasten up and control the polymerisation process, while yet others such as surfactants, colours and others are added to achieve pre-designed properties in the PU.

4.1.1 Polyols

In polymer chemistry, polyols are compounds with multiple hydroxyl functional groups available for organic reaction. They are high molecular weight materials manufactured with an initiator and monomeric building blocks. Polyols are generally classified into polyether polyols and polyester polyols. Polyether polyols are made by the reaction of epoxides (oxiranes) with active hydrogen as the starter compound. Simple epoxides are named from their parent compounds, ethylene oxide or oxirane, such as in chloromethyloxirane.

Polymerisation of an epoxide gives a polyether. For example, ethylene oxide polymerises to give polyethylene oxide. Polyester polyols are made by poly-condensation of multifunctional carbolic acids and hydroxyl compounds. Polyether polyols are probably the more commonly used ones as they offer technical and commercial advantages such as low cost, ease of handling and better hydraulic stability over polyester polyols.

https://doi.org/10.1515/9783110643169-004

Polyols can be further classified according to their end use as flexible or rigid polyols, depending on molecular weights and their functionality. Flexible polyols will have molecular weights from 2,000 to 10,000 with OH number ranging from 18 to 56. Rigid polyols have molecular weights from 250 to 700 with OH numbers ranging from 300 to 700. Polyols with molecular weights from 700 to 2,000 and OH numbers ranging from 60 to 280 are used to add stiffness as well as increase solubility of low molecular weight glycols in high molecular weight polyols.

4.1.2 Isocyanates

An isocyanate is a functional group with a chemical formula $-N=C=O$. Organic compounds that contain an isocyanate group are referred to as *isocyanates*. An isocyanate that has two isocyanate groups is called a *diisocyanate*. Isocyanates with two or more functional groups are required for cycloaliphatic isocyanates.

Aromatic isocyanates, which are the most produced all over the world are always the preferred ones for manufacturing PUs. The main reason for this is that they are much more reactive than the aliphatic ones and more economical to use. The two most important commercial and popular isocyanates are: *toluene diisocyanate* (TDI) and *diphenylemethane diisocyanate* (MDI).

TDI is a colourless pale yellow liquid that has a density of 1.214 g/cu.cm with a melting point of 21.80 °C (71.24 °F) and a boiling point of 2510 °C (4550 °F). With a slightly pungent smell, it is a low viscosity toxic liquid. TDI exists in two isomers, namely: 2, 4 –TDI and 2, 6-TDI and is commercially produced and marketed as either 80/20 (TD-80) or 65/35 (TD-65) and always a mixture of both isomers.

MDI on the other hand, exists in three isomers: (1) 2, 2-MDI, (2) 2, 4-MDI and (3) 4, 4-MDI but the 4, 4 isomer is most widely used and also known as pure MDI and is also used in the manufacture of PU products.

4.1.3 Water (primary blowing agent)

Water is the primarily blowing agent for PU foams. The polymerisation reaction between a polyol and an isocyanate in itself will not produce any gaseous products. This reaction alone will only produce a solid, extremely high density, rigid and hard mass of PU. What gives PU foam its low mass to volume ratio is the expansion of the PU polymer being formed. Though liquid gases like carbon dioxide can be introduced into the reaction mix to achieve a foaming effect, it is far easier for endogenously generated carbon dioxide to be more economical and the foaming easier to control. To achieve this, water is blended into the reaction mix. The exothermic reaction between water and the isocyanate will generate carbon dioxide and urea. A combination of the heat generated from the polymerisation and from the water/isocyanate reaction makes the carbon dioxide gas to expand within the yet soft polymer being formed, thus increasing the volume of the polymer. In most flexible foam productions, water is the primarily blowing agent.

Generally, local municipal water supplies are good enough but water with high concentrations of dissolved or suspended metals are not acceptable, unless treated. This is because unwanted metals in the water may interfere with the catalysts used in a PU formula.

4.1.4 Auxiliary blowing agents

An auxiliary blowing agent (e.g., methylene chloride) is used in combination with water to produce low density foams, less than 21 kg/cu.m or to produce soft foams at all densities. These auxiliary blowing agents are generally liquids with low boiling points.

4.1.5 Silicone surfactants

Surfactants are usually organic compounds that are amphiphilic, meaning they have both hydrophilic and hydrophobic groups. Silicone surfactants are organo-modified branched silicone polymers with many 'heads' and 'tails'. Their tails are hydrophilic, while their heads are hydrophobic and this makes them more efficient than other surfactants and cost-effective to use. When making PUR foams, surfactants not only emulsify, disperse and help in the high speed stabilisation of foam cells during the expansion period but also later. There are many classes of silicone surfactants but the majority of flexible foams are made with a class of surfactants identified as *polysiloxane–polyoxyalkylene copolymers*, which are viscous, glassy, slippery and slightly dark liquids.

The polyoxyalkylene portion of the surfactant assists in solubilisation of the surfactant into the polyol and then helps in overall emulsification. The polysiloxane portion of the surfactant lowers the bulk surface tension. This process is called stabilisation and the silicone surfactant in foaming is known as a foam stabiliser, in simple terms.

4.1.6 Catalyst systems

A catalyst is a substance that alters the rate of a chemical reaction but remains unchanged. While positive catalysts increase or accelerate, negative catalysts will decrease the rate of a chemical reaction. In PU foaming all catalysts are positive ones. Virtually all commercial foams are made with the aid of at least one catalyst but for producing good quality foams various combinations of catalysts are used.

These systems used in very minute quantities in a formulation mix will establish an optimum balance between the chain propagation and the blowing (isocyanate/water) reaction. The type and concentration of catalysts can be selected as per process requirements such as cream time, rising profile, gel time and even curing of the outer surface skin. There are two broad classes of catalysts used in PUR foam production. They are amine catalysts and the organometallic catalysts, commonly known as tin catalysts.

4.1.6.1 Amine catalysts
Amine catalysts like triethylenediamine (TEDA) with the trade name DABCO and dimethylethanolamine (DMEA) are some of the common amines used for the production of PUs. Tertiary amines are generally selected on the basis of whether they drive the reaction to some extent or whether they are further selected based on how much they favour one reaction over the other. For example, tetramethyl butanediediamine (TMBDA) drives the gelling reaction better than the blowing reaction. On the other hand, TEDA or DABCO will drive the blowing reaction better than the gelling reaction.

4.1.6.2 Organometallic catalysts
An organometallic compound is a compound containing at least one metal-to-carbon bond in which the carbon is part of an organic group. They are used as catalysts and as intermediates in industrial productions. Organometallic compounds based on mercury, lead, bismuth, tin and zinc can all be used as catalysts in PUR productions. The polymer forming reaction (gelling) between an isocyanate and polyol is promoted by organometallic catalysts, with probably tin compounds being most popular, depending on the type of production.

4.1.7 Basic additives

Apart from producing a standard product, to obtain special qualities or properties in PUs, certain basic additives are required. Whatever the end use of a PUR product, the first basic need is its aesthetic value for consumers, especially for products that are sold uncovered, meaning it has to be coloured. Other properties required may be load bearing capacity, fire resistance, resilience, mechanical and physical properties and so on and to meet these requirements additives are required in the formulations.

4.1.7.1 Colourants
PUR foams when produced are 'white' in colour and most manufacturers use various colours colour codes for easy identification of different densities and grades.

A basic yellow is the colour used to counter UV action which turns the foam into a light brown colour because of degradation. Suitable pigments or dyes can be introduced into the polyol for thorough blending before other components are mixed in. Typical inorganic colouring agents include titanium dioxide (white/shades), iron oxides and chromium oxide. In two-component systems, a producer has a choice of custom-order of a coloured system or adding a colour to Part B, which is generally the blended polyol/polyol system with other components. A good practice would be to dissolve a weighed amount of colour with a small amount of the polyol and mixing it well, before introducing it into the main body of the polyol system. Where colour suppliers have a code, for example from 0 to 8, the higher numbers will give the best colouring and colour fastness.

4.1.7.2 Fillers

Fillers generally used are very fine particle size inorganic compounds, added to PUR formulations to increase density, load bearing, sound absorption and reducing costs. Of all the range of many fillers available, the most used one is calcium carbonate which blends well with polyol systems. Fillers should be as dry as possible with minimum moisture content as the primary blowing agent is water and there are two possible problems with increased water levels. One is that the final texture of the foam and density will be affected and the other is a possible fire hazard, because of the exothermic (heat giving) reaction taking place, if the water content in the formulation is over the threshold. If the moisture content is high in the filler, a producer may want to make an adjustment to the calculated water content intended as the blowing agent. Then again, if the moisture content is very low and negligible, an adjustment may not be needed.

4.1.7.3 Flame retardants

PUR open-celled foams with low density has large surface areas and high permeability of possible ignition and it is a common practice to include a flame retardant to reduce the possibility of flammability. The choice of flame retardant for any specific foam often depends on the end flammability that may be influenced by additives that include the initial ignition possibility, rate of burning and smoke evolution. The most widely used flame retardants are the chlorinated phosphate esters, chlorinated paraffin and melamine powders.

4.1.7.4 Graft polyols

Graft polyether polyols contain copolymerised styrene acrylonitrile with low-inhibition compounds designed to maximise load bearing properties and will form a polyol system as desired depending on the end applications. These special polymers can have a solid content up to 45% and will also produce cost-effective foams.

4.1.7.5 Extenders

An extender is a special type of polyol of many short chains which connects the longer chains of the base polyol or blends. Chain extenders/cross-linkers are usually low-molecular weight polyols or polyamines that are used to improve the properties of PUR foams by aiding the curing process.

4.2 Packaging for two-component systems

In two-component systems, a producer will receive fully blended systems of their choice. This choice will naturally depend on the end application and easy to process as the two-components will be made up of component A (isocyanate TDI or MDI) and component B (polyol/polyols +additives). Component B, the polyol blend will generally contain a blended mixture of polyol, chain extenders, fillers, catalysts, surfactants and blowing agents with colour being an option. The suppliers of these systems will provide technical information as well as the mixing ratios of A and B and make it easy for a PU products manufacturer for processing. On request, suppliers will include additional additives to counter flammability, UV, microbial action and others as needed in component B.

Different suppliers may have their own types of containers but the general packaging systems for two-component systems are as follows:

- Small packs in small plastic containers like 20 or 50 litres or in small steel drums. For small spray applications a user can use a standard spray gun, while for in-situ building spray applications larger spraying devices will have to be used and the larger packs will probably be preferred.
- One of the most common practices is for steel drums, painted in two colours like blue and red, green and red or other to identify each component. Preferred colour for component A (isocyanate) can be red. These drums will have two bung holes with lids, one smaller and the other larger. The bung hole with the smaller lid should be opened slowly to let off any pressure build-up inside, while a long-stem stirrer can be inserted through the larger opening and the contents in component B should be mixed well before removal of contents for use.
- Totes: These are for large volume packs and there are several different types like steel, large plastic container in metal mesh, plastic containers encased in cardboard. These are suitable for processing through machines, where components A and B are connected to a dispensing machine and the pre-mixed material is introduced into a mould. In these systems, the mix ratio and exact volume are pre-set. Totes are also called bulk drums because they hold large volumes. Figure 4.1(a, b) and shows some packs.

(a)

(b)

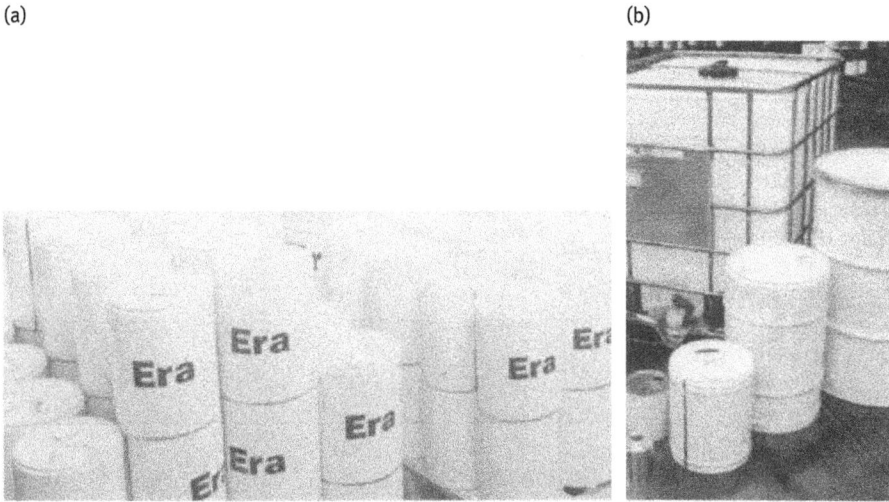

Figure 4.1 (a) and (b): Two-component packs.
Source: Reproduced with permission from ERA Polymers Pty. Ltd. (Australia)

4.3 Non-traditional biomass fillers

Fillers are an essential part of polymer production technology. Although fillers may affect the quality of PU products in one way or the other, they are required to achieve properties like load bearing, support factor, cost-effectiveness and so on. Although calcium carbonate is the most used filler for PU foams, there are others that are used for commercial polymers and for PU composites. Table 4.1 shows some of these commonly used ones.

With growing environmental concerns, the polymer industry has been seeking ways to increase the uses of biomass as substitutes for traditional materials for some time now. Because of constant research the last few years based on emerging technologies have revealed some interesting new possibilities for fillers for polymers such as rice hull flour, rice hull ash, walnut shell powder, wheat hull flour, bamboo flour, bamboo fibre, expandable polystyrene (EPS), calcium carbonate (from egg shells), graphene (for increased thermal conductivity) and so on.

Out of these, probably the most important two are graphenes and bamboo. As examples, while graphene has tremendous strength-giving and thermal conducting properties ideal for automotive applications, bamboo fibres/polymer combinations are now producing excellent soft textiles and the bamboo fibres can easily replace fibreglass reinforcements in epoxy/fibre composites such as boat building and similar applications. In the solar energy industry, it has been found that graphene-coated substrates even with much thinner coatings give better energy absorption and retention properties than ones coated with cadmium or other. Table 4.2 shows some non-traditional fillers.

Table 4.1: Some fillers used in commercial polymers.

Inorganic	Organic
Glass	Carbon
Calcium carbonate	Polymers
Iron oxide	Wood flour
Magnesium carbonate	Cellulose
Titanium dioxide	Wool
Zinc oxide	Aramid fibre
Zirconia	Nylons
Hydrated alumina	Polyesters
Antimony oxide	–
Metal powder	–
Silica	–
Clays	–
Barium ferrite	–
Silicon carbide	–
Potassium titanate	–

Table 4.2: Non-traditional fillers.

Item	Form	Function
Rice hulls	Flour	Filler/stiffening agent
Rice hulls	Ash	Moisture barrier/filler
Wheat hulls	Flour	Composite filler
Walnut shells	Flour	Composite filler/foams
Egg shells	Powder	Filler ($CaCO_3$)
Bamboo	Fibre	Composites (reinforcement)
Bamboo	Flour/powder	Filler/composites
Graphene	Powder	Foam-special properties
Expandable polystyrene	Beads	Foams

Fillers can constitute either a major or a minor part of a PU composite. Considering their relative inherent higher stiffness properties compared to the polymer matrix, they will always modify the mechanical properties of the final-filled products or composites. The structures of filler particles can range from precise geometrical forms, such as spheres, hexagonal plates or short or long fibres to irregular masses. Based on 100 parts of polymer by weight or volume, the recommended maximum is 100 parts of filler, where the mixture becomes a thick slurry. Some technologies may want to go beyond this reasonable limit up to 150 parts but at this level it will be almost impossible to handle the mixture for pouring and so on. Highly filled PUs are generally classed as 'cheap' foams where properties are not important.

Fillers can also be classified according to their source, function, composition or morphology. No single classification may be adequate due to the overlap and ambiguity of these categories. Generally, fillers are used in a PU formulation as a single item but producers can also use filler systems in order to get the maximum benefit of the wide range of compatible fillers available with different properties like stiffness, strength, dimensional stability, toughness, heat distortion, damping, impermeability and cost reduction.

4.4 Handling raw materials safely in brief

The safety factor of using two-component systems on a factory floor is much less than all chemicals being purchased in bulk and handling them. Since the two-components systems are delivered to a factory floor in sealed containers, the first step in safety handling begins with the proper risk-free unloading under supervision. Since both components are flammable and corrosive, maximum safety handling should be observed. The 'handlers' should wear protective clothing, goggles and other and ensure that no sparks are created in the case of steel drums. All personnel should be familiar with the standards as stated in MSDS (safety data). Before the materials are taken for use, the pressure build-up inside each container should be neutralised by opening the smaller bung slowly then closing it tightly. The larger bung lid can now be opened and the material inside stirred slowly by inserting a long-handle mixing device. This is especially important for the container having the polyol blend, since there is a possibility of small solids having been formed.

Large volume producers who have in-house laboratories, may use the option of taking random samples to check against the invoices/technical information provided by the material suppliers but they being professional organisations can be relied on for quality supplies. Here, a quick test – the 'box test' or a 'cup test' can confirm the quality or the recommended ratios by the supplier. The next step would be the drawing of small quantities from each component, sufficient for small volume productions. Here, there could be spills, if not handled properly and the supervisors or a lead-hand will have to employ spill-management techniques. If any of the handlers come in contact with either of the two materials, especially the isocyanate, it should be immediately washed with water thoroughly. Suitable absorbents will have to be used for mopping up the spilled liquids. For larger productions, the whole drums/containers can be transported carefully to the dispensing machine area and connected to it.

For all PU production operations, for the protection of personnel basic safety wear plus safety equipment is a must, in addition to eyewash stations, showers, first aid stations, fire extinguishers, emergency exists and so on. For small volume producers, at least eyewash stations and a first aid station should be available on the factory floor. For all operations, periodic checks for air quality from outside services would greatly help in protecting the health of all operational personnel. Training in spill management, safety factors, fire drill and so on, are areas which will help a trouble-free and efficient operation. Disposal of the empty drums or smaller containers should be according to guidelines as per MSDS but more in keeping with local municipal disposal regulations as these drums can be termed as hazardous. The big volume plastic totes can probably be returned to the suppliers for re-use.

4.5 Managing spills

When setting up a PU production operation, whether it is a small volume or large volume operation, it is best to have a thorough knowledge of the hazardous nature of the chemicals being used. There are many safety standards and technical information available from different sources to assist and organisation on how best to ensure proper transport, unloading, storage and use of the corrosive and flammable materials.

Isocyanates – vapour inhalation is the first problem followed by possible spills on a production floor. Spills occur mostly when handling material for weighing, mixing, pouring or also form leaking containers. Spills and leaks should be immediately contained and cleaned by trained personnel. In case of a spill or leaks, all personnel in the area should be evacuated immediately. Standard protective wear should include self-contained breathing device, as well as protective clothing, footwear and gloves and safety glasses. An approved respiratory device must be worn if there is a possibility of isocyanate vapours exceeding the recommended threshold limit of 0.05 ppm (time-weighted average) limit set by the Occupational Safety & Health Administration (OSHA).

Polyols – If the spill is small on the factory floor, it can be absorbed with sawdust, rice hulls, wheat hulls or other absorbents available and then swept up and disposed. Large spills should be pumped into containers and then disposed. Personnel engaged in this operation must wear at least eye and skin protection as well as protective footwear.

In cleanup procedures, there are standard neutralising liquids and a suitable one consists of 5% aqueous ammonia (sodium carbonate) and 1–2% detergent in water. Isocyanate spills will need more attention and thorough cleaning unlike polyols and final decontamination may be achieved by spraying the area with large quantities of water, only after thorough removal of all traces of isocyanate.

4.6 Raw material storage

In a normal large volume PU production set up where the raw materials are purchased in bulk, an organisation would have a separate building or at least a section away from the production area to be able to have more control of safe storage due to the inherent possibilities of the raw materials. Although contact with polyols will only probably cause irritation, which can be effectively washed off, the isocyanates are high-risk hazardous materials and calls for maximum caution in storage and handling. For two-component systems, although the two materials are in secure steel or plastic containers and may not need a separate storage building as such, nevertheless careful and effective storage facilities are needed.

The size of the area needed will naturally depend on the quantity of materials in stock at a time. This area can be minimised if purchases are done on a just-in-time (JIT) basis and should be a cool place with provision for storing the two components separately with easy access for removal and also to the production area.

The polyol side of a typical system is typically not as hazardous as the isocyanate side. However, caution is recommended. Formulated and blended polyol systems often contain volatile agents or catalysts such as amines, which may pose hazards under certain conditions. If these systems contain low boiling chemicals, they must not be stored at temperatures that may cause pressure build-up inside and cause container rupture. The isocyanate side of a system is hazardous and flammable. Different systems may use different isocyanates. The most common ones are based on TDI or MDI which may be present as pre-polymers, polymeric or crude materials and should be treated with equal respect. Inhalation is the most common cause for exposure, so it is advisable to be aware of the vapour pressure of the products, when opening the lids for use. Using at elevated temperatures, will increase the vapour hazard.

Bibliography

1. Era Polymers Pty. Ltd. – Polyurethane Systems – www.erapol.com.au
2. FOAM-TECH: Urethane Foam-Open cell vs. Closed cell – www.foam-tech.com
3. Flexible Polyurethane Foam: Raw Materials – mychemicalengineer.com
4. Dow Hyperlast Advance: an update in Polymer Engineering – www.hyperlast.com
5. Defonseka, Chris – Practical Guide to Flexible Polyurethane Foams – pages 145–148 – Smithers Rapra 2013

5 Mould designs

5.1 Introduction

This chapter will present an important aspect of processing polyurethanes (PUs) – *mould designs*. The processing of two-component PUR systems will require moulds of varying types, from simple designs to more complex ones for moulded foam and for the continuous foaming systems. This last category will not be discussed as we are dealing with two-component systems and foam producers who deal in large volume productions and will use continuous foaming systems, where the raw materials are connected to a mixing-head and fed from multi-streaming tanks. Since the overall subject is two-component PU systems, it is more practical to base discussions on small-to-medium volume productions from mould designs to fabrication of suitable moulds and final processing methods.

The author with actual hands-on foaming and mould designing experience will present in detail, innovative and cost-effective ways of designing and making suitable moulds for (a) hand-mixed productions, (b) standard single moulding productions and (c) large single block productions. Although large PUR foam blocks are made by professional dispensing machines using an *intermittent process* – since the machines are expensive, a later chapter Design of Foaming Plant and moulds will discuss the designing of a simple foaming plant using innovative technology. Here, only the designing and fabrication of suitable moulds will be discussed for two-component PU systems.

When designing a mould, one must consider the interaction of five important parameters – the end product, material to be used, processing method, desired end properties including aesthetic values and the effective design of a mould to encompass all these. Computer-assisted mould making like computer-aided designing and computer-assisted machining (CAD/CAM) can be used to design the more complicated moulds. These computer systems are especially useful in designing contoured or multi-cavity moulds. All moulds will have precise tolerances acceptable to whatever standards are being used or as per custom orders. Modern day technologies allow the manufacture of any type of mould, single or multi-cavity, intricate patterns or others, not only at reasonable costs but with high aesthetic values.

For moulded parts, a producer may opt for the manufacture of a prototype for which cheaper materials can be used for making the mould and for sample approval by a customer before investing in a final mould. In general, a good mould will have a lifespan around 20,000 to 50,000 mould cycles. Some may last longer, depending on the material used and the skills applied by a mould maker. Some of the important aspects of a mould are as follows: product geometry, weight, shrinkage, venting, closing/opening system, engraving, plating for desired finish (smooth, rough, matt and glossy), provision for inserts, cooling system, clamping system, ejector system and so on. If dispensing machines are used for processing

https://doi.org/10.1515/9783110643169-005

the mould or moulds, they must be compatible with the machine for easy operation, whereas if the PUR mixture is to be poured into an open mould, it is much simpler.

Here, we will examine the possibilities of using cost-effective materials like wood, fibreglass and some others for making moulds for two-component productions, except for moulds for moulded items like integral skin products, where the processing methods will require steel moulds. With a good understanding of two-component PU systems, it is a big advantage, especially for an entrepreneur or small producers to be able to use cost-effective materials for moulds for reduction in production costs.

5.2 Design parameters

While the desired end properties of a product will be as per the materials formulated, the quality of a final product will greatly depend on the design and the quality of the mould. Therefore, before embarking on designing a mould, one should have a thorough knowledge of the product to be manufactured. Moulds for open-pour processes will be much easier than closed mould productions and the design requirements will be somewhat different. Designing moulds can be very exciting but challenging at the same time. For example, the final quality requirements of a product made for consumers will be quite different to the high quality standards of a product made for industrial or automotive purposes.

In the case of moulded products, the process becomes a little complicated with certain aspects of a moulding cycle like – filling, clamping and mould – heating, dwell time, cooling, part removal and so on being the main operations. When designing moulds, areas that need special attention are material quality, injecting of correct material volume, material seepage, proper heating and cooling, positioning of inserts (if any), sufficient clamping force, residual stresses, shrinkage, warping, required density, aesthetic values or other. Safety factors and moulding cycle times are also important for efficient and cost-effective production. The simple wooden mould design recommended should be able to accommodate most two-component PUR systems and the size can be varied accordingly to produce the dimensions of any final product desired.

When designing moulds for processing PUR two-component systems, it is important to consider the interactions between some of the basic areas shown below to ensure quality products.
- Marketable value
- Aesthetic values
- End application requirements
- Dimensional requirements
- Tolerances
- Processing factors

- Mould costs
- Mould life
- Product costs

5.3 High-tech mould designing for integral skin products

Design and engineering are keys to building better moulds. In high-end and sophis-ticated markets for products for the automobile, aircraft, space travel, hotel in-dustries and so on, the need for precise, aesthetically pleasing, high quality and cost-effective products is paramount and this calls for high quality moulds with the shortest lead times. When products with intricate patterns and super surface fin-ishes are needed, the mould takes top priority, designs have progressed over the years and the latest software technologies are utilising mould making much easier, with greater precision, cost-effectiveness and shorter delivery times.

In present days, machining moulds without a right software is like a builder try-ing to build a housing complex without a blueprint. The latest development in the software industry is by Autodesk Canada Company, which has brought together – Delcam, HSM, Netfabb and Majestic Systems as a portfolio of solutions for software manufacturing, covering modelling. CAM, additive manufacturing, composites, robotics fabrications, inspection and factory layout among other functions. Auto-desk's cloud-based product innovation platform – which combines CAD/CAM and CAE in a single package allows users to take their designs all the way to production with 3D printing capabilities and HSM works for multiple-axis milling machines, turning centres and waterjets.

Some of the other software tools are Vericut 8.0 from CGTech (USA), Missler Software (France) and others, all highly useful for efficient production of any type and quality of moulds. 3D Systems has released Version 13 of its Cimatron software, featuring a broad range of new CAD for tooling functionalities for faster design, including direct modelling.

Having high-tech tooling software is one thing but selection of suitable tool steel for turning out a mould or moulds is also an essential factor. The aesthetic values of the moulded surfaces will call for specialised work and if intricate pat-terns are involved, the work needed to achieve them will need more skills. When suitable tool steels are chosen, a mould maker will take into account the life span of a mould as required by a customer, which generally means the number of accept-able mouldings that can be achieved before resurfacing or minor repairs have to be affected or the mould rejected totally.

There are many specialised mould makers and also companies that produce special tool steels to meet the demand for moulding high-tech products. One such is SCHMOLZ + BICKENBACH (US and Canada) with over 160 years of steel produc-tion experience backing their brands is not only a pioneer but market leader in

speciality steels. The use of the latest technology has enabled them to meet the toughest requirements in tool steels in terms of:
- Degree of purity and polishability
- Uniformity of hardness and microstructure
- Wear and temperature resistance
- Machinability, toughness and hardness
- Thermal conductivity

Their specialised tool steel materials for moulds are: Formadur grades: 2083 Superclean, 2085, 2311, 2316 and Corroplast FM: P20 Modified, P20 High Hard, P20 ESR, Holder Block, S7 Mould Quality, 2344 Superclean (H13 ESR) and Moldmax grades.

5.4 Hand-mixed productions

This is the simplest form of producing PU products using two-component systems. There are many ready-made (standard) or custom-made two-component PUR system suppliers who will provide the two components as: component A (isocyanate) and component B (blended polyol) which is the norm for these systems. Both will be in liquid form and should be de-pressurised and mixed thoroughly before use.

5.4.1 Mould design for hand-mixed pouring

This is the simplest form of a mould and cost-effective method of making flexible PU foam cushions for furniture applications using a two-component system. It can be used to make either standard or high-resilience foams, which will depend on the material being poured into the mould. This type of manufacture will be for small volumes (around 2,000 cushions/month), with low capital investment, where the required moulds, a simple cutting system can be fabricated on a production floor and the raw materials can even be purchased on a just-in-time (JIT) basis. High quality can be achieved with good processing methods. Ideal for a small volume producer for direct supply to a furniture manufacturer or as a contractor and also for an entrepreneur. Here, only the making of a mould is presented as detailed processing methods will be discussed in a later chapter 5 (5.4.1.1).

Raw material system: Premilec Maxflex 421 for moulded flexible foams. This is a fully formulated two-component flexible foam PUR system using water as the blowing agent. It is designed for production of furniture foam products, using either open or closed mould pouring techniques. Although, relevant specific processing parameters are given below, it is always advisable to carry out a 'box-test' (small sample) to adjust them, if necessary, to achieve premium quality.

Components properties

	Component A	Component B
Viscosity @25 °C (77 °F)	210 cps	400–600 cps
Specific gravity	1.19	1.03
Mixing ratio by weight	50	100
Cream time (s)	12–16	–
Gel time (s)	95–105	–
De-mould time (minutes)	7–10	–
Free-rise density	4.5 lbs. /cu. ft. 72.0 kg/cu.m	–

Product: Foam cushions for furniture
Sizes: 20 inches (50 cm) × 20 inches (50 cm) × 4 inches (10 cm)

The first problem to counter is when the material in liquid form is mixed and poured into the mould, to prevent seepage from the base of the mould. The average viscosities of the mixture will give an indication of the fluidity. The usual practice would be to use a gasket at the base but this can cause problems and also be expensive, as the gaskets will be affected by the PUR mix needing frequent replacements. The author would suggest a simpler solution in that the mould should 'sit' on a polythene sheet covered piece of flexible foam. The weight of the wooden mould will make the mould 'sink' into the soft base and will prevent any leaks. The foam sheets trimmed from the blocks can be used for this.

From the material specifications provided previously, an approximate moulding cycle time can be worked out which will indicate the number of moulds required to meet pre-set production targets. For example, each moulding is expected to yield 10 cushions.

Quantity/block: 10 cushions

The material for making the mould can be wood, laminated board or aluminium. For the sake of saving costs, wood will be used, which is good enough. Some of the important aspects when designing this mould are material leakage, pressure build-up inside, inside surfaces to be smooth as possible to prevent sticking, prevention of meniscus (curved top), foam block shrinkage, easy de-moulding system, mould dimensions and so on. The mould will comprise two parts – (1) the main mould and (2) a floating lid.

5.4.1.1 The mould

Although the mould is a simple wooden one, it must be solid, strong and warp-free. To construct a mould use 3–4 cm thick wood with detachable sides and bottom. The bottom base should have a groove to accommodate the assembled vertical sides of the mould, which should fit together and prevent leaks, with the real problem being at the bottom. As the foam rises it is in a soft solid state and the chances of material leaking is minimal. However, clamping is advised to negate the pressure build-up inside the mould due to the exothermic chemical reaction taking place inside. The inner dimensions of the mould when assembled should be 51 cm × 51 cm (allowance for block shrinkage and trimming of sides) and the internal height of the mould should be around 120 cm plus. This is because, when cutting the fully cured foam block, small amounts of material will be lost due to the thicknesses of the cutting wire/blade. If hot-wire cutting is used (not cutting wire due to burning), conventional slitting machines are used, and allowance have to be made for the thickness of the blade. For example, you cannot cut 10 cushions, each 10 cm thick from a well-trimmed foam block of 50 cm × 50 cm × 100 cm. The result will be nine cushions plus one cushion less than 10 cm thick. This factor can be overcome by adjusting the volume/weight of the pour but the mould must be able to accommodate the additional height needed.

5.4.1.2 The floating lid

When the PUR mix is poured into the mould, as the foam rises, when it has reached about one-third of the mould height, a meniscus (surface curve) will start to form. Unless prevented, this material will be a waste. To prevent this, a 'floating lid' made up of a very light material, for example, plywood – 50 cm × 50 cm with a handle in the middle is needed to place on top of the rising foam.

While closed-mould processing will require venting to allow gas buildup to escape, in this open-pour simple wooden moulds, the gases have ample room to escape from the wide open top. For non-adherence of foam to bottom and sides, a good release agent can be used. A spray would be the easiest and best. Some producers may opt to use thin polythene sheeting for lining the sides of a wooden mould but this can cause problems due to the material creases, which can mean thicker trimmings to reach a smooth surface. When constructing the four wooden panels to form a box to vertically fit into a bottom base, a mould designer may decide on a 'tongue and groove' arrangement or other. Also, if the mould base is mounted on wheels it will greatly help in mobility. Here, innovation will be a great advantage. Figure 5.1 shows some MDF boards which are strong and have a smooth laminated surface.

Figure 5.1: MDF boards.
Source: Adapted from Polymer Composites by Hardy Smith Ltd., India.

5.4.2 Moulds for larger rigid foam blocks for taxidermy

Taxidermists have been relying on PUR foams for creating beautiful figures of animals, fish or others. This is a high-end and highly skillful industry needing just the right grade of foam for carving as the basics of building a particular figure. For this purpose large foam blocks are needed and this section deals with the fabrication of a simple mould from wood. Other materials that can be used are metal or aluminium.

Consider the production of a fully trimmed foam block of size – 36 inches (90 cm) × 30 inches (75 cm) × 30 inches (75 cm) height. The recommended thickness of the wood panels to be used is about 3–4 cm with the internal dimensions of the assembled wooden mould – 91 cm × 76 cm × 31 cm (with allowances for shrinkage and trimming). The general principle of the design can be on the same lines as recommended in Section 5.3.1 with detachable sides and the mould base on wheels for easy movement. The size of the mould can vary as per desired final size of foam block and even though the volume is fairly large, still two-component systems can be easily used.

To calculate the volume/size of foam block required use the the following formula:

$V = M/D$ where, Volume = length × width × height M = mass or weight D = density

Or, if only volume calculation is needed, use $V = L × W × H$

Make allowances for shrinkage, trimming and 1% for material loss due to gas.

5.4.2.1 Example of raw material system
Erathane TX56 – a rigid two-component PU system from ERA Polymers Ltd. (Australia) is specially designed for taxidermy applications. When mixed in the correct recommended ratio, it will produce a foam with a free-rise density of 56 kg/cu.m.

Component properties

	Polyol	Isocyanate
Appearance	Clear, honey coloured liquid	Brown liquid
Viscosity (cps)	300	250
Specific gravity	1.15	1.22
Mixing ratio	100	100
Mixing time (seconds)	20	–
Cream time (seconds)	120	–
Gel time (seconds)	270	–
Tack-free time (seconds)	420	–
Free-rise density (kg/cu. m)	56	–

Here, the tack-free time can be taken as the de-moulding time and will give an indication of the number of actual moulds required. The viscosity factor will indicate the liquidity factor which will be helpful in preventing material leaks, especially at the initial pour time.

5.4.3 Moulds for circular flexible foam blocks

Foam sheets of varying densities and thicknesses are widely used in bedding, furniture, footwear, packaging, clothing and other industries. The bedding industry in particular needs these foam sheets in wide widths and in continuous form for mattress padding and quilting. These sheets generally range from densities 0.9 lbs/cu ft. (approximately 14 kg/cu.m) to higher densities as per individual mattress manufacturer's requirements. They are mostly lower density range foams, very flexible and pliable but required in wide widths to be able to use as padding even for Queen and King size mattresses. These continuous sheeting is cut (peeled) from circular foam blocks of varying diameters and lengths from around 36 inches (90 cm) to 96 inches plus (240 cm) and since the slitting blades can be adjusted, varying thicknesses can be cut. Probably the maximum length (height of the foam block) will be about 36 inches (90 cm) for hand-mixed pour systems and larger volumes will require dispensing machines.

The first step is to produce a large circular foam block which is then cut into thin sheets called – *peeling* – using a peeling machine, which slits the foam block into thin continuous sheeting rolls. The usual thicknesses used in the mattress industry for padding is around 2 to 25 mm but this can vary. For peeling sheets, the foam should be soft, pliable and have uniform density across the width of the foam roll, when placed horizontally on the peeling machine. Coarse foam textures will

create problems by adhering to the slitting blades and giving damaged surfaces of foam due to the heat being generated by the slitting blades. Flexible PU foam sheets are the preferred material for this application due to the following properties:
- Compressibility
- Cushioning
- Flexibility
- Lightweight
- Mildew resistant
- Resiliency
- Vibration damping

Moulds can be made out of aluminium or fibreglass or any other suitable material with a base and a standard floating lid. These finished circular blocks have to be mounted horizontally on a metal bar of a definite diameter on peeling machines. This means the foam block must have a suitable 'built-in core'. The bottom base of the circular mould must have a round groove to match the diameter of the horizontal bar of the peeling machine to accommodate a vertical core, which can be a strong tube of plastics, cardboard, metal or other, which can be put in place before the PUR mixture is poured in. The design and placement of this core should be such that it will not move or be displaced during the foam rising period and be able to be part of the circular foam block when de-moulded. Alternatively, solid circular foam blocks can be made and then using a boring machine, the desired bore-hole can be drilled to insert a hollow mandrel. Some manufacturers may opt to cut sheets from square blocks but this will mean extra waste.

To calculate the pour volume use the required formulae,

$$V = \pi r2 \cdot h \text{ where } V = \text{volume of circular block, } r = \text{radius and } h = \text{height}$$

Mass = Volume × Density. This is to calculate the weight ratio of components. Make allowances for shrinkage, trims and 1% loss of material due to gas. Figure 5.2 shows a peeling machine where the circular foam rolls are 'peeled' into continuous sheets.

5.4.4 Moulds for high resilience PUR foam mattresses

Although, foam mattresses are made from standard flexible PUR foams, manufacturers may want to make them for high-resilience foams for high-end grade mattresses. These will be naturally heavier and can be made in a soft or harder texture depending on the formulations. Again, wood can be used as a mould material (3–4 cm thick) but the inside height of the mould can be 42 inches (105 cm) to achieve a moulded block height of 40 inches (100 cm).

Since the pour volume is large, it is advisable to dispense through a PUR dispensing machine. Using the standard formula,

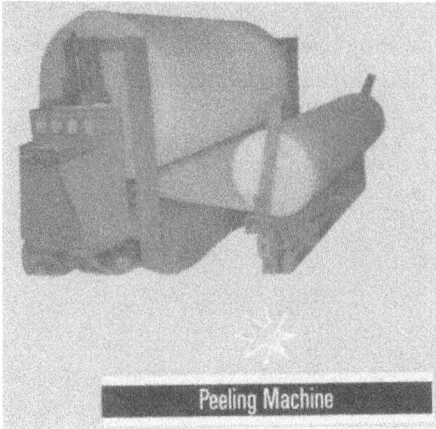

Figure 5.2: A peeling machine.
Source: Reproduced with permission from Modern Enterprises, India.

$$M = V \times D \text{ where } M = \text{weight of pour } V = \text{volume } D = \text{density}$$

A suitable dispensing machine can be worked out. As shown earlier, a complete production cycle time will indicate the number of moulds required against the desired production volumes.

5.4.5 Moulds for viscoelastic (memory foam) mattresses

Because of the high viscosity of the foam mix and especially the high moulded foam density, foam blocks are moulded in a maximum height of 28 inches (70 cm). The recommended height is 24 inches (60 cm) but 28 inches is possible. Generally, good memory foams will range from 48–96 kg/cu.m with mattresses with higher densities, graded as high-end quality. For making a mould, a producer may opt for the one made with aluminium sheets instead of wood for easier handling and smoother sides, although the costs would be more than that for wood. These costs are however justified as the final product would sell for 3–4 times a normal foam mattress.

The aluminium sheets should be strong and not warp because of the pressure build-up inside the mould from the exothermic reaction taking place. The use of aluminium will help in much less adherence of material to the mould sides due to the smoothness of aluminium and cost savings in using lesser amounts of a release agent or none at all. The mould construction with a solid wood/aluminium combination is also a possibility. De-moulding time will be longer than for standard foams and also the post-cure period will be a minimum of 48 hours instead of 24 hours. When selecting a suitable dispensing machine in addition to the shot or pour weight, the viscosity of the mix should also be a factor.

5.5 Standard single moulding productions

The two-component systems are ideal for moulding small or large individual items. Moulds can be designed with intricate patterns to hold inserts, high gloss or matt effects, self-skinning surfaces (although this will depend more on the material than the mould surface) and so on. Single mouldings are required by sectors such as consumers, industrial, automotive, space travel, aircraft industries and many others. Probably the more important and largest volume of moulded PUR products are footwear, office chairs, aircraft and auto seats, auto accessories, sports goods and many others.

Take the case of manufacturing automotive accessories; for example, an integral skin PUR foam seat. This should be moulded in one piece according to specifications supplied by the customer. These productions will be needed in large volume productions and a manufacturer will probably opt for a number of moulds made of fibre glass material with a pour hole and adequate venting. A PUR dispensing machine connected to the two-components can deliver the mix via a mixing head and fill the closed moulds through the pour holes. If metal moulds are used with different moulding systems, the moulds will have longer lives but involves much higher capital costs and also give better 'skin' finishes. There are two methods of making these integral skin moulded products – (1) heat cured and (2) cold cured, with the latter being the easier and less costlier method, depending on the two-component material grade system being used.

With the latest technologies available to a foam producer, products with pleasing aesthetic values and great quality can be achieved, including patterned or printed skin surfaces. Some advantages of these mouldings are thermal conductivity foam, inserts for exterior attachments and so on, and it is advisable to invest in high-tech moulds and suitable processing machinery, especially when dealing with products for the automobile and aircraft industries where strict industrial standards have to be met. Figure 5.3 shows a moulded integral skin PUR seat.

The first step is to determine the product requirements before starting to design a mould to produce this specific product. The load functionality, tolerances, aesthetic values, weight, industrial specifications, environmental requirements and others will determine the size, thickness of skin, surface finish and foaming system to be used. The overall cost factor will also play a key role. Once the overall functional requirements have been worked out a mould designer can generate a 3D CAD model which will help to visualise the final product and possible problems. Modification of this model may be needed to finalise an acceptable final model on which the mould maker can proceed.

Figure 5.3: PUR integral skin moulded seat.

These foams generate pressure on a mould. Therefore, it is important to incorporate a few small (about 1 mm) vent holes in the mould which will allow air and gas to escape. It is however important to allow only a minimal amount of material to escape. If a large amount escapes either through these vent holes or the part-lines due to insufficient clamping force, large holes will be created near these escape points. The vent holes should be kept to a minimum and not too large to prevent foam shrinkage or surface imperfections of the moulded foam. Figure 5.4 shows a microcellular sole which is moulded in two halves using a steel mould.

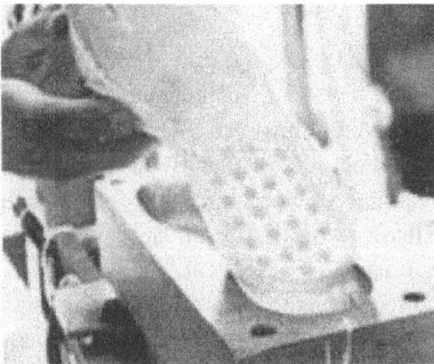

Figure 5.4: A microcellular sole.
Source: Reproduced with permission from ERA Polymers Ltd., Australia.

When using a new mould, some surface conditioning may be needed. It is recommended that 2 to 3 coats of release agent be applied, one after the other, with sufficient time between each coat to allow the solvent to evaporate. A heavy duty wax-based release agent with a fast drying solvent base is ideal. Different surface textures may be achieved by spraying different mould release agents. The recommended mould temperature should be at 35–40 °C (95–104 °F) to achieve moulded products with a smooth silky finish on the surface.

5.5.1 Raw material system (for auto seats)

Ecofoam ISF136 is an Integral Skin Polyurethane Foam product with a free-rise density of 135 kg/cu.m and environmentally friendly foam that has no ozone depleting potential. This system can be easily pigmented and can be manually drilled mixed at a minimum speed of 2,000 rpm for small mouldings; however, for large mouldings it should be processed through a plural PU dispensing machine.

Component properties

	Polyol –component B	Isocyanate-component A
Appearance	Hazy white liquid	Brown liquid
Viscosity (cps) @ 20 °C	725	145
Specific gravity	1.07	1.2
Mix ratio by weight	100	37
Mix time (seconds)	10	–
Cream time (seconds)	24	–
Gel time (seconds)	49	–
Tack-free time (seconds)	75	–
Free-rise density (kg/cu.m)	135	–

This grade of material is made by ERA Polymers Ltd. (Australia). They also represent manufacturers of PUR dispensing machines and can also supply custom-made two-component systems to suit different products. When selecting a dispensing machine, important points to remember are mould compatibility, minimum/maximum shot dispensing capacity and mixing speed compatibility.

5.6 Moulds for large size foam block productions

Generally, large size PUR foam blocks are made either by a continuous foaming system and cut into blocks or by an intermittent process where each block is moulded one at a time. These processes use multiple streams of raw materials through a single mixing head before pour or dispensing. They are ideal for very large volume productions. However, it is possible to use two-component systems also for the intermittent processes, where the two components, blended polyol (component B) and isocyanate (component A) in large drums or totes are connected to a dispensing machine and poured/dispensed through a mixing head into a pre-determined large mould.

In two-component systems only two items are to be accurately measured or weighed as against measuring or weighing of multiple components, and the risk element of errors is much higher. Moulds made from wood would be good enough so long as they are strong and warp-free. The use of two-component PUR systems for block manufacture is especially suitable for a small producer with limited knowledge of polymer chemistry/PUs or an entrepreneur with limited starting capital.

5.6.1 Recommended mould fabrication

Moulds can be of any size and generally will be only limited to the dispensing machine to be used or already available. All dispensing machines will have a maximum shot dispensing weight and this will have to be a little more than the full weight of the PUR mix to be dispensed. A machine must dispense the full amount of material needed as one shot or pour to achieve good quality foam.

5.6.2 Example

A foam manufacturer wishes to make foam blocks from which he can cut foam slabs to be used as standard single size mattresses. He decides on a common market size of – 76 in. (190 cm) × 36 in. (90 cm) × 4 in. (10 cm) per mattress. He plans a production cycle of 10 mattresses from each foam block and making allowances for shrinkage, trims, material loss due to gas and an allowance for material loss due to band saw cutting, he works out that a moulded foam block of size – 77.5 in. × 37.5 in. × 42 in. (height) is needed.

He designs a large rectangular mould made of wood, with all four vertical sides detachable and the base mounted on wheels for easy post-moulding movement. A floating lid made of very light plywood of size – 77 in. × 37 in. × 0.25 in. is also designed and fabricated, with two handles for easy handling. The inner dimensions of the mould is – 77.5 in. × 37.5 in. × 45 in. (height). If necessary, drill a few very small holes (1 mm diameter) at strategic points on the floating lid to allow gas generated during foaming to escape. Innovative designs can use cheap wood lined with thin aluminium sheets or even warp-free rigid plastics sheets or others to ensure a smooth surface to prevent the foam sticking on to the surfaces. Polythene film lining, Kraft paper lining, spray coats and other methods can be used but could be bothersome and also costly.

5.6.3 Final product specifications

- 77 5 in. × 37.5 in × 42 in. foam block = 70.64 cu.ft.
- Required density = 2.0 lbs/cu.ft.
- Raw material profile: Rise time 145 seconds, de-mould time 10 minutes

Then, applying formula, $M = V \times D$ where M= weight V= volume D = density

$$M = 70.64 \times 2.0 \text{ lbs/cu.ft.} = 141.28 \text{ lbs.} = 64.22 \text{ kg} + 3\% \text{ (gas loss)} = 66.15 \text{ kg}$$

Therefore, this mould is suitable for use with a PUR dispensing machine having a minimum pour weight or shot weight of around 70 kg or more. Taking into account the de-moulding time plus other production time factors, the number of moulds required for a particular volume of production can be worked out. Generally, it is advisable to have at least two moulds for alternate moulding cycles to minimise the production downtime.

Bibliography

1. ERA Polymers Ltd. (Australia) – 'Moulded Foams' –mixing procedures – 11/4/2007
2. ERA Polymers Ltd. (Australia) – Two-Component Systems – mould fabrication – 11/4/2007
3. Premilec Inc. – 'Flexible Moulded Foam Systems' – Mould Fabrication Guidelines
4. Oswald-Bauer-Brinkmann – 'International Plastics Handbook': Engineering Design-pages 451–476 -2006 Hanser Publications Munich
5. Defonseka, Chris – 'Practical Guide to Flexible Polyurethane Foams' – pages 71–81 2013 Smithers Rapra UK
6. SoftwareModern Enterprises India – www.foam-machinery.com
7. Autodesk Inc. – Engineering Software
8. Hardy Smith Ltd. India – WPC Composite Technology
9. Schmolz + Bickenbach Group – Canada: Tool Steel Solutions
10. Cimatron E (Israel) – CAD/CAM Solutions for Tool Making

6 Selection of machinery

6.1 The concept of processing two-component systems

The concept of processing two-component polyurethane (PU) systems available as liquids – component A and component B – involves accurate weighing or metering of each component and mixing together to arrive at a homogenous mix. The actual production process will consist of three phases – (1) where the components are mixed in their containers itself before use, (2) laboratory tests to check out workable production parameters and (3) the production process itself. All three stages will require different types of machinery and equipment. The methods of dispensing this mix into moulds will be on the principles of – *open pour, high- pressure or low-pressure dispensing systems, injection, casting* or *potting*. The first four will be presented in detail as they are the most common processes and the last two to a lesser degree.

Although high-quality two-component PU fully blended material systems are available from specialised suppliers or through their agents, to obtain high-quality final products, it is essential to understand the need to use efficient machinery and equipment to process these systems. This can range from manual, semi-automatic to fully automatic operations. This discussion will be confined to manual and semi-auto processing as the popular processing methods as fully auto systems are multiple component streaming systems and will probably remain so due to its flexibility to produce different densities and foams of different qualities.

The concept of two-component PU systems is that as a general rule, component A will contain an isocyanate, while component B will contain polyol blended with the rest of the ingredients of a formulation. These systems will be able to produce only a standard density and properties as formulated for but in some of the systems the densities could be varied within reasonable limits by varying the ratios of the two components. These systems can be available as coloured or without coloured, as a foam producer may want to make different colours of foam which can be easily done on a factory floor.

Naturally these PU systems will have a specific shelf-life, generally around six months and for optimum results, both components, especially component B should be thoroughly mixed in the drum itself before use.

6.2 Selection of equipment for pre-production processing

Although the concept of just mixing two components of chemicals for foaming seems simple enough, there is a correct procedure to follow. In order to have a trouble-free operation and produce good quality foams and foam products, the following sequence of processing is recommended:

https://doi.org/10.1515/9783110643169-006

- Mixing both components in the drum/container before use.
- Quality assurance – pre-production 'cup test' or 'box test' in a laboratory
- Accurate weighing/metering of each component
- Correct mixing speeds as recommended by the material supplier
- Accurate mixing time before pour, injection or dispensing
- De-moulding after sufficient cure time – non-sticky

6.2.1 Pre-production material mixing (phase 1)

Generally, PU two-component systems are supplied in plastic drums (small volumes), steel drums (larger volumes) and even in plastic totes for very large volumes. Although, these containers will be stored at temperatures as recommended by the raw material suppliers, in a factory environment, the temperatures can fluctuate and affect both the contents and the containers. Because of the nature of the liquids, coagulation can take place with slight lumping and hinder free-flow when pumping to the mixing-head. The plastic containers can 'expand' due to the build-up of internal pressure. To counter these, periodic action by slowly opening the smaller 'bung' and letting out the internal pressure and closing the lid should be carried out. The larger 'bung' should be opened and the liquid contents of both components should be mixed slowly for a short period by inserting a hand-held motorised drum mixer with a long shaft with a mixing disc. It is essential that the lid is closed back tightly to prevent the entry of any moisture, which can alter the pre-formulated density and also cause the possibility of fire if the water content of the polyol blend exceeds the threshold limit of maximum water content in a formula.

6.2.2 Pre-production 'Cup-Test' or 'Box-Test' (phase 2)

It is recommended that all productions of PUs, whether small or large volume operations have an in-house laboratory. Even a basic set up to check pre-production tests for foam quality and density would suffice. Large volume producers may even have equipment to test each component of A and B for quality but generally it is not required as the foam producers can purchase their needs from reputed suppliers who will also supply detailed specifications and also provide full operational data.
  The basic equipment needed would be:
- A small electronic weighing machine – 1 g to 100 g
- Few glass beakers with volume markings
- One or two small glass mixing vessels
- One or two glass rods to be used as stirrers
- Hand-held high speed small mixer
- Mould release agent – silicone-based or other

- Two small hand-pumps to fit the larger bung of the containers
- Two small chemical-resistant taps to fit larger bung of the containers
- A wooden box – 30 cm × 30 cm × 15 cm height with detachable sides
- A table model small hot-wire machine (optional)
- Safety equipment like gloves, aprons, eye protectors
- Others as required

6.2.2.1 Cup-test

Weigh the mixing vessel to be used and small quantities of components A and B in the ratio as recommended by the system supplier in separate containers. For example, ratios of 100:30 or 100:100 or other. First, put component B (polyol blend) into the mixing vessel and then pour component A (isocyanate). Using the glass rod or hand-held mixer stir mix for a few seconds. The mix will start to react and slowly turn a 'cream colour' and start to foam and rise slowly and may well overflow forming a 'mushroom' effect. Allow to cool, until foam is not tacky and then remove foam from the mixing vessel and weigh the foam mass.

6.2.2.2 Box test

This test can be more accurate as the volume needed to fill the box can be calculated and the pour-in liquid volume can be fairly accurate.

For example, Volume = $30 \times 30 \times 15$ = 13,500 cu. cm = 0.0135 cu. m
Weight = $V \times$ Density (35 kg/cu. m) = 0.473 Kg.

Now, weigh the two components separately in the given ratios. Apply release agents to the inside surfaces of the box. Pour component B into the mixing vessel and then component A and mix in speed for a few seconds. Pour the mix still in the liquid state quickly into the box (before it starts 'creaming') and allow the foam to rise slowly. A slight curve will form at the surface. Allow the foam mass to cool and when the foam is not tacky, de-mould and remove the foam block.

From these tests, a foam producer can work out valuable information before the actual production starts to avoid, material loss, density adjustments, colour requirements, production parameters and so on. Remember, a raw material system supplier will supply technical and processing data as recommendations only which has to be adjusted to suit a producer's working conditions in relation to floor temperatures, equipment used and other factors.

Tests will indicate:
- Cream time (initiation time)
- Gel time
- Free-rise time
- Top of cup fill time

- Meniscus-forming time (surface curve)
- Tack-free time
- De-moulding time
- Free-rise density
- Density versus ratios
- Material loss due to gas losses
- Foam shrinkage factor

More detailed procedure for these tests can be had from industrial standards ASTM D7487 (2013) for Standard Practice for Polyurethane Raw Materials: Foam Cup Test.

6.3 Machinery for mixing and dispensing (phase 3)

The processing of two-component PU systems can be done from a simple innovative manual method, semi-automatic or fully automatic feed systems for multiple moulds.

6.3.1 Manual mixing and pouring

This is the simplest form of producing PUR foams and when the raw material systems are in two component form, it is that much easy. This method is ideal for making foam cushions, blocks, wedges, sheets for garment padding, medical needs like post-surgical seats, wheelchair cushions and others.

The four important areas are weighing, mixing, pouring into mould and fabrication. The basic equipment needed would be as follows:

Example:
To make foam cushions of size: 50 cm × 50 cm × 10 cm (20 in. x 20 in. × 4 in.)
(a) Machinery: A hand-held single-phase electric drill with adjustable speeds up to 3,000 rpm with forward and reverse functions. A steel rod approximately 1.25–1.50 cm in diameterand 70–75 cm in length, with a 10–12 cm round mixing disc with four flanges on one side attached to the end of the rod. The other end should be firmly inserted and locked into simple, high-speed hand-held 'mixing device' that should suffice as a mixing device for a two-component system, where a supplier of these systems specifies *drill-mixed* or *machine-mixed*.
(b) Equipment: Basic equipment would be:
    - 0–1 kg and 0–10 kg electronic weighing machines
    - 4–5 large plastics buckets with handles
    - Two drum stirrers
    - Two manual or electric pumps

- Two horizontal drum stands (optional)
- Two chemical corrosion-resistant taps for drums
- One or two drum trolleys
- Two platform trolleys
- Miscellaneous tools
- Safety wear

Moulds: These can be made from wood, aluminium or other similar materials. The number of moulds will depend on the volume of production and the shapes and sizes will depend on the types and sizes of foam blocks to be made. To construct a simple mould, use 4 cm wood with detachable sides and the bottom base of the mould should have four grooves to accommodate the four vertical side panels which should fit tight to prevent leakages. The real problem comes at the bottom plate to prevent material leakage as the PUR mix will be in liquid form when poured into the mould. Chemical resistant seals may be used but they may have to be replaced from time to time. For the above-sized cushions, the sides of the inner dimensions of the mould when assembled should be 51 cm × 51 cm × H (ht. as per requirement). A light 'floating lid' of dimensions 50 cm × 50 cm and very thin thickness (e.g. plywood) with a small handle on the centre of the lid will be required to prevent the meniscus being formed by the rising foam.

6.3.1.1 Cutting and fabrication

The foam blocks after a minimum of 24 hours curing will need thee trimming from all sides. This involves horizontal and vertical cutting and a band saw or a hotwire cutting machine can be used. While the latter will give a very smooth cut and surface with negligible material loss due to very thin thickness of the wire, a band saw will also give an adequate cut and surface but the material loss with each cut will have to be provided, as the high-speed cutting blade will be much thicker in diameter.

Some countries may not allow hotwire cutting because of possible fire hazards, which is very unlikely. Band saw machines are available with manual, semi-automatic or fully automatic operations and will cost more. For the benefit of a small foam producer or an entrepreneur, the author shows how a simple hotwire cutting machine will be cost-effective and can be made on the factory floor.

6.3.1.2 Hot-wire cutting machine

These machines can be purchased from many sources in some countries. One such company is –Demand Products USA which has a wide range of these machines and also accessories. The basic materials required to make an effective cutting machine for the foam blocks under review are as follows:

- Single-phase 15 amp or 30 Voltage Controller – 10 to 100 V variable
- Nickel–Chrome 14 g or 16 gauge wire

- Set of electrical clips(single wire) with spring attachments
- Two lengths of aluminium channel – 150 cm × 5 cm/1 cm
- A 240 cm × 120 cm × 2.5 cm (thick) warp-free board
- A steel or metal frame with legs
- A slow-speed motorised device with forward/reverse motion
- Miscellaneous tools

Assembly basics would be to mount the board on the steel frame horizontally or at 45° angle for gravity feed (if motor is not used). Cut grooves or one long groove in the aluminium channel for upward and downward movement of the hotwire arrangement and fix them on either side of frame. Attach the hotwire arrangement via spring load and clips. Connect the two electrical power cords coming from the power controller to the two aluminium verticals. Fix a measured length of Ni–Cr wire across the table to the two aluminium arms. Turn on the power controller and note the voltage at which the foam cuts easily giving a smooth surface. If desired, attach multi-wires to increase the production output, in which case use the 30 A power controller.

6.3.1.3 Machine for foam waste recycling
For small volume operations a three-phase shredder would suffice. Here, a machine with the correct set of blades as recommended by the supplier should be carefully selected as it is difficult to shred foam wastes because of its softness and pliability. A large volume foam producer may have several shredders and also a re-bonding machine where the foam pieces after shredding are compressed into foam blocks, which in turn are cut into sheeting for carpet underlay, mattress bases and so on.

6.3.2 Low-pressure dispensing machines

The dispensing machines presented here for both low pressure and high pressure is to give the reader an idea of the machinery only and it should be noted that there are many designs and types offered by a wide range of suppliers in many countries.

Low-pressure machines are most often selected for low-volume productions. Mixing of the chemicals takes place in a chamber, which includes an electric motor-driven or hydraulically operated dynamic mixer resembling an auger with multiple grooves. This dynamic mixer can have a speed range from 3,000–8,000 rpm depending on the recommendations of the material systems suppliers. Piston-type pneumatically operated chemical injectors retract to allow the blending of compo-nents at pressure ranging from 30 to 500 psi. When the required 'shot' of blended mix is dispensed, the injectors close. After a desired production run, the mix head is usually flushed with the solvent or water and compressed air to purge any remaining chemicals from the mixing chamber.

There are many different types of low-pressure machines from many suppliers but all are based on the same principle where the two components are mixed to pre-set ratios and mixing times and dispensed through a mix-head. Some machines may have a fixed dispensing weight/volume but more advanced machines will have better functions. The low-pressure dispensing machines shown in Figure 6.1 are versatile units capable of dispensing from 1.5 kg–100 kg per minute.

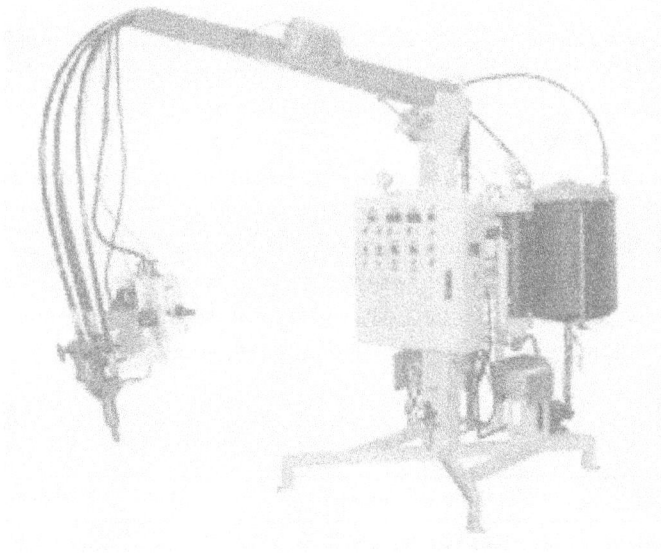

Figure 6.1: A low-pressure dispensing machine.
Source: Reproduced with permission from Polycraftpuf Machines Ltd., India.

These LP-series low-pressure PU foam dispensing machines are available with an environmentally friendly, hot water flush capable and circulating-type mix head. The mix head and mixing chamber has the option of either adjustable hydraulic or variable speed electric mixer. Polycraft LP series low-pressure machines offer an economical and environmentally clean alternative to the industry for processing PU foams and elastomers, adhesives and so on for a broad range of applications.

LP-series is powered by a Programmable Logic Control (PLC) control and operator process control with display of the working parameters. LP-series output and ratio variation are powered by a Variable Frequency Drive (VFD) drive with manual flow control of components.

Technical specifications

Model	Ratio A:B	Ratio variation A:B	Output (kg/min)	Tank Capacities A/B	Power (kW)	Weight (kg)	Size
LP 15	1:1	1.5/5.1	1.5–15	50 litres each	7.5	575	3,300×3,000×1,100
LP 30	1:1	1.5/5.1	5–30	100 litres each	10.0	650	3,700×3,000×1,500
LP 60	1:1	1.5/5.1	15–60	150 litres each	15.0	750	3,700×3,000×1,500
LP100	1:1	1.5/5.1	22–100	150 litres each	15.0	850	4,000×3,000×1,800

6.3.3 High-pressure dispensing machines

High-pressure dispensing machines for PUs are generally used in applications that require measured shots to produce moulded foams. The principle involves the blending of chemical components as formulated under high pressure in a mixing chamber of the mix head. A hydraulically controlled plunger in the mix head retracts, allowing two or more chemicals to merge together under high pressure in the range of 1,200 psi to 3,000 psi in the chamber. After metering a precise shot of the mix, the plunger closes and retracts and cleans the mix head automatically and does not require flushing. In this case under review, there will be only two streams: components A and B. The following technical data is a typical standard high-pressure dispensing machine as supplied by Trad(e) Belt Company, South Korea.

TB standard models

Model	Output g/cc	Metering motor	Pump output	Tank capacity
TB H50	50–100	3.75 kW/800 rpm	2 cc/rev.	250L +250L
TB H100	100–300	5.5 kW/800 rpm	6 cc/rev.	250L+250L
TB H 500	300–800	7.5 kW/800 rpm	11 cc/rev.	300L+300L
TB H 1000	800–1500	15 kW/1100 rpm	33 cc/rev.	500L+500L
TB H 2000	1500–2000	22 kW/1100 rpm	62 cc/rev.	1000L+1000L
TB H 3000	2500–3500	37 kW/1100 rpm	92 cc/rev.	1500L+1500L
TB H 5000	3500–6000	45 kW/1100 rpm	125 cc/rev.	2000L+2000L

Products

The following are a few of the items made using these machines:

Building materials: sandwich panels, sliding panel, heat retardant door and door sealing

Automotive parts: seat cushion, instrument panel, head rest, arm rest, bumpers and console box

Miscellaneous: air cleaner, filter, packaging, ski goods, surfboards and seats

6.3.4 Multiple products production

The two common processing methods are: manual operation and fully automatic moulding operations.

6.3.4.1 Manual moulding operation
Example:
Production of PUR soles and shoes. Figure 6.2 shows some of the items that can be produced using two-component PU systems.

Figure 6.2: Items produced using the two-component polyurethane systems.
Source: Reproduced with permission from Polycraftpuf Machines Pvt. Ltd., India.

Polycraftpuf Machines Pvt. Ltd., India specialises in manufacturing PU footwear and accessories using two-component systems. The simplest unit for these productions consists of two basic elements. A low-pressure open-pour dispensing unit in which the mixer is attached to the end of a boom that can be moved from mould to mould by an operator. Multiple moulds mounted on a circular flatbed will move in a circulatory movement with adjustable speeds. For small volume productions, the opening and closing of each mould can be done by a single operator.

This plant can accommodate 24–30 moulds, with an hourly output range from 120 to 150 pairs (240–300 pieces) minimum, depending on the machine model. Figure 6.3 shows the machine and Table 6.1 shows the machine specifications.

Figure 6.3: Large volume automatic moulding machine.
Source: Reproduced with permission from Polycraftpuf Machines Pvt. Ltd., India.

Table 6.1: Machine specifications.

	Machine specifications
Variable ratio	1:5–5:1
Jacketed tanks	200 kg
Solvent tank	20 kg
Boom	@ 180 degree swing
Control	PLC with colour animated panel
Machine power	10 KW
Number of moulds	24–30
Conveyor power	24 KW
Heating system	Individual mould heating

This versatile machine can produce products of different densities, colours and properties as per formulated components.

6.3.4.2 Fully automatic moulding operation

The HC-201 series High Pressure from HuizhouHeChen Mechanical Equipment Co. Ltd., China is a good example of a fully automatic moulding operation for two-component PUR systems (table 6.2). This machine consists of a hydraulic, pneumatic and automatic frequency conversion systems, electronic control system, mass metering system, magnetic control, temperature control, pre-mixing system and a PLC central control system as main functions.

Technical specifications

Table 6.2: Machine parameters.

	Machine parameters
Material	Polyol/Isocyanate
Size	370 L × 200 W × 250 H (cm)
Weight	2500 kg
Total power	380 V 50 Hz
Total power	55 kW
Energy consumption	40 kW
Touch screen	22.5 cm × 17 cm colour display

This versatile machinery system can accommodate up to 60 moulds for PUR shoe soles or similar products and can produce around 4,000 pairs of soles per day. This high-pressure dispensing machine is ideal for making memory foam pillows, safety shoes, leisure shoes, automobile parts, ski accessories, computer parts, office equipment parts and many others.

6.3.5 Injection dispensing

Instead of presenting one machine and one process, the author for the benefit of the readers will present an overall view of the fast developing processing methodologies leading up to the current advanced practices. PU processing equipment have shown incremental advances in cost-effectiveness and the ability to handle a wide range of applications, which includes new mixing and metering machines capable of handling abrasives, highly filled formulations, composites, viscoelastic foams and very small or very large output rates. Of main interest here is that of reaction injection moulding (RIM) as an especially dramatic moulding process that combines RIM and thermoplastic injection moulding in one mould and machinery system.

6.3.5.1 RIM plus injection moulding

One of the new developments in PUR processing by Krauss–Maffei is the connection of a 2,530 ton MX thermoplastic injection machine to a RIM-Star Compact 16/8 RIM system and four MicroDos colour metering units. This mammoth system produced a inner door panel for cars with a polycarbonate/acrylonitrile (PC/ABS) substrate over-moulded with three colours of PUR.

The injection machine moulds the substrate on one side of a rotating centre cube tool. Then the mould opens and the cube rotates 180° and while it is open, a robot arm sprays mould release on the RIM side of the tool. After the mould closes, another PC/ABS substrate is moulded on one through three heads mounted on the moving half of the mould. Another robot arm will de-mould the finished product.

6.3.5.2 Highly filled PUR hybrids

Because of growing demands for highly filled and reinforced PUR formulations, two firms: (1) Krauss–Maffei and (2) Canon have made new 'hybrid' dosing systems by using piston metering for the abrasive polyol side and conventional pump metering of the isocyanate. Canon has revised its 'HF' hybrid design, making it more standardised and economical as well as more compact. The close-loop control of both components has been updated, detecting the actual values of the two flows by measuring the linear displacement of the dosing piston. Canon calls this system a 'plug-and-play' machine that is ready to go, right out of the box.

Another system for highly filled PUR formulations comes from Impianti OMS S.p.A in Italy. This company uses a volumetric gear pump specially adapted to handle high levels of mineral Eco-fillers blending and metering unit designed for recycling PUR foam scrap that is ground into powder and mixed into the polyol. This system includes a day tank that mixes and temperature conditions of the polyol/powder slurry. The slurry is metered by a gear pump to the mix head as a third component to the unfilled polyol and isocyanate components.

6.3.5.3 New spraying hardware

Cannon offers a machine that can spray high-density rigid PUR foam, either reinforced or unreinforced, as backup for the thermoformed plastic skins in bathtubs, showers, spas, campers and boats. The two components are sprayed by Cannon's new AG6 high-pressure metering unit which uses rotary pumps. This system can also spray polyurea gelcoats directly onto a mould.

Hennecke GmbH has a machine that can spray-mould PUR composites. In this system, one head can dispense layers of different materials that may be filled, unfilled or glass-reinforced. A new development is the MN 6 CSM high-pressure mix head for small outputs. It is especially useful for applying thin layers or for access to constrained spaces in a mould. The spray volume reduction can be from 40 g/s down to 6 g/s. Hennecke's largest high-pressure head is the new two-component MT 36, designed for 500–5000 g/s. Three of these heads can fill a 13.5-metre long insulated truck in one step.

6.3.6 Machinery for polyurethanes as potting materials

This presentation will be on two-component PUR system for potting and encapsulating material for microelectronics. EPIC RESINS, a leading electronic potting compound manufacturer offers a variety of PU formulations to suit different applications designed to protect sensitive electronic parts and similar parts from environmental and chemical stresses.

PUs cover a broad range of materials and can replicate any type of hardness for products including gel, glass, thermoplastics, rubber and LED compounds. EPIC PU formulations can be available to meet various end properties such as chemical resistance, weather damage, excellent insulating properties, suppress vibration in varying thicknesses, viscosities and other requirements depending on the end applications and industry.

For example, EPIC 7356 is a two-component PU potting and encapsulating compound specially used for automotive sensors, switches and other devices requiring 'under the hood' environments. This range of potting compounds or systems has fast gel and curing times, which means increased production cycles. As for machinery for this operation, there are models from manual desktop to table models to fully automatic machines for very large volumes. Details of two models are given below based on simple production methods:

1. **Auto desktop polyurethane potting machine**

   Supplier: Dongguan Yiren Industry Co. Ltd., China
   Model: No. Y4D 7400N
   Applications: auto parts, sensors, cellphones, buttons, LCD sealant, circuit boards and other

2. **Dispensing machine for potting polyurethanes**

Supplier: Ashby Cross Company UK
Model: DXBG Bench Top Model
Description: ideal for use with urethanes for applications such as potting, protective coating, sealing and so on.

Bibliography

1. Polycraftpuf Machine Pvt. Ltd. – India – www.polycraftput.com/pu_lpunit.htm
2. Naitove, Mathew.H – article 'PUR Machinery Gains in Versatility and Economy' – 2008 – Plastics Technology magazine
3. Epic Resins – Resins for Microelectronics – www.epicresins.com/Electronics Potting/microelectronics
4. HuizhouHeChen Mechanical Equipment Co. Ltd. – China – automatic casting machines – www.burrillandco.com

7 Calculations, formulating and formulations for polyurethanes

Generally, a polyurethane (PU) foam manufacturer must have a thorough knowledge of all components that combine to produce a quality foam. Although only three basic chemicals – (1) polyol, (2) an isocyanate and (3) water combines to form a PUR foam, certain additives have to be added to promote, control and achieve pre-determined properties. Two of the main properties, which may be considered important can – (1) foam density and (2) foam cell structure.

Since, this presentation is about two-component systems, where all components are pre-blended into two chemicals and easy to foam, the foaming process is simpler with the mixing of two components A and B – in ratios as recommended by the system suppliers under certain temperature conditions. We must also be aware that other factors such as actual temperatures at the time of processing, the machinery and equipment used and processing techniques used will also have a bearing on the quality of the final product. However, for the benefit of the readers and technologists, the author will present in detail – calculations, formulating techniques, some basic formulations and some operational financial aspects in the hope that this information will be useful.

7.1 Calculations

7.1.1 Density

In PUR foaming, density is one of the key factors of a quality foam, and density can be defined as the mass per unit volume. This calculation is denoted by the formula:

$M = V \times D$ where M = mass (weight) in lbs. or kg
V = volume in inches or cm
D = density in lbs/cu ft. or kg/cu.m

Example:
What is the density of a foam mattress of size – 80 in. length – 60 in. width × 6 in. height and weighing 35 lbs?

Applying, $M = V \times D$ 35 lbs. = 16.67 cu ft. × D
Therefore, D = 2.1 lbs/cu ft. or 33.6 kg/cu m

https://doi.org/10.1515/9783110643169-007

7.1.2 Indentation force deflection

The indentation force deflection (IFD) is basically the support factor of a foam, whether it be for a cushion, mattress or otheris a valuable marketing tool. It is the ratio of compression of a selected representative sample foam when compressed to 25% and 65% of its height, using an electromechanical device with a standard size plate. The values of the forces required to compress the foam are noted and then the ratio of 65%:25% is the IFD value. If this value is ≥ 2.0, the foam quality is acceptable and any value below 2.0 is considered below standard. If this occurs, foam producers will solve it by slightly increasing the filler content in the formulation to increase the load bearing factor. The support factor is also known as the *compression modulus*.

$$\text{Support factor} = 65\% \text{ (IFD)} \div 25\% \text{ (IFD)}$$

7.1.3 Load bearing capacities for polyurethanes

Unlike flexible PUR foams, moulded PUR foams have excellent load bearing abilities. These high load bearing capacities allow moulded urethane components to be made thinner or smaller resulting in reduced weight and cost of raw materials. The magnitude of compressive force which can be supported by a given PU compound is dependent on three main factors: (1) hardness, (2) loaded surface condition and (3) shape factor.

7.1.3.1 Hardness
Hardness is defined as the relative resistance of a surface to indentation by an indenter of specified dimensions under a specific load. Hardness is generally measured with the use of an indenter. A harder PU sample will be more resistant to indentation by the indenter and thus, naturally, will give a higher durometer reading.

The hardness range of elastomers is so broad that a single durometer cannot indicate practical measurable differences of hardness. For this reason, durometers are available in more than one type and are used to measure the hardness of hard PUs. For comparison purposes, a rubber band is about 35A, the tire thread in a car is about 70A and the wheels of a roller skate are about 85A and a hard hat is about 75D. The hardness factor can be useful for designing engineers, especially in the auto industry, when designing auto parts.

When a PU part is deflected, any factor that influences the part's ability to bulge will affect the stress–strain relationship. One such factor is the loading surface conditions. If a PU part was compressed between two rigid parallel plates, the PU surfaces in contact with the plates would tend to spread out laterally under a load. If the surfaces were lubricated, this spreading will occur much easier than if the plates were clean and dry. If the urethane part was bonded to the plates, then this movement would be completely restricted. In this case, the total area of part that is free to move is reduced, thus effectively stiffening the part.

7.1.3.2 Shape factor

Shape factor is a term used to mathematically describe the shape of a part. It is equal to the area of one loaded surface divided by the total area free to bulge. Two parts, regardless of the size, that are made of the same compound and have the same shape factor will behave similarly when loaded in compression. This is extremely helpful to the design engineer because large parts can be scaled down for easy testing in laboratory conditions. It is also very important to note that the shape factor relationship is only plausible when the loaded surfaces of the PU parts are completely constricted from spreading out under the applied load. For loading conditions where the loaded surfaces are not constrained, load-deflection tests are required to fully understand the load–deflection relationship.

The shape factor for common load bearing applications will fall somewhere between 0.25 and 2.0. When shape factor falls below 0.25, the part will be at a high risk of buckling. Much the same way a column in a building would buckle, if it were made too long and slender. When the shape factor gets much higher than 2.0, the effective compression modulus approaches the bulk loaded surface area.

The following example shows how a design engineer can use the concept of shape factor in conjunction with compression stress–strain curves to estimate the load-deflection characteristics for PU parts.

Example

A commercial truck manufacturer is currently using PU bumpers between the cab and frame to isolate vibration and shock loading. The bumper is 3 in. (7.5 cm) in diameter and 1 in. (2.5 cm) thick. A new luxury cab design to be introduced to the market but according to the new design the added weight will overload the PU bumpers. The truck maker needs to accommodate this added weight; however, he cannot alter the size or shape of the standard bumper due to existing mounting brackets. He would also like to stay with the same compound because there is not sufficient time to fully test and validate a new compound. What can he do to accommodate this added weight? (See Figure 7.1.)

a) The shape factor of current bumper:

$$\text{Shape factor} = \text{Loaded area} \div \text{Bulge area}$$
$$= (\pi \times 1.5 \times 1.5) \div (\pi \times 3 \times 1) = 0.75$$

b) In order to accommodate a heavier weight a higher shape factor than 0.75 needs to be used. In order to raise the shape factor without changing the overall design, a sandwich bumper can be made by bonding a rigid plate in the middle which will increase the overall stiffness. (See Figure 7.2.)

c) Shape factor of new sandwich design will be:

$$\text{Shape factor} = \text{Loaded area} \div \text{Bulge area}$$
$$= (\pi \times 1.5 \times 1.5) \div (\pi \times 3 \times 5) = 1.5$$

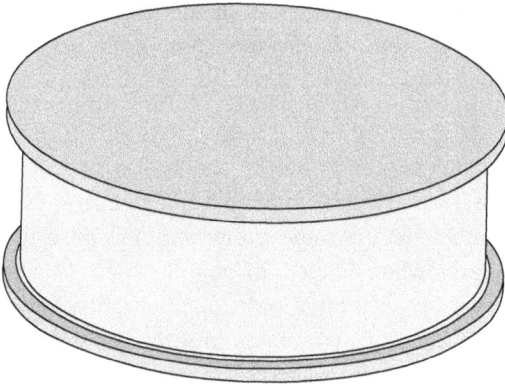

Figure 7.1: Polyurethane Bumper.
Source: Image curtesy of Gallagher Knowledge Center.

Figure 7.2: Reinforced Polyurethane Bumper.
Source: Image curtesy of Gallagher Knowledge Center.

d) Now the overall deflection using the new shape factor needs to be multiplied by two to account for the two bumpers in series. The new overall deflection, however, will be less than the deflection based on the original design.

7.2 Cell structures of polyurethanes

PU foams can be formulated to have a range of cell structures from soft to semi-rigid to semi-rigid. This property is closely associated with the density of a foam, which in turn can be varied to suit end applications. For example, soft foams are needed for comfort applications, whereas semi-rigid and rigid foams are used for moulded products, insulation and special applications. Most people when buying comfort products like mattresses or cushions will look for popular brand names, density and the support factor. For insulation applications, the important aspects are the R-value (thermal conductivity factor), ease of application and durability. In practice, rarely would the subject of 'cell structures' arise, as neither the buyer nor the seller would know much about it, although to a foam producer it would be very important.

Open-cell foams are soft with the cell walls or the surfaces of the bubbles broken and air filling all the spaces of the material. This makes the foam soft or weak but with stability. The insulation value of the foam is related to insulation value of the calm air inside the matrix of broken cells. Closed-cell foams vary in degree of hardness, depending on density. Most of the bubbles or cells in the foam are not broken, representing a field of inflated balloons or soccer balls piled together in a compact configuration. This makes it strong or rigid because the stabilised bubbles or cells are strong enough to take a lot of pressure. One can walk on some rigid foam without much distortion. It is possible also to fill these cells with a special gas to increase the normally efficient R-value.

The advantage of closed-cell foams compared to open-cell foams include its strength, higher R-value and greater resistance to the leakage of air or water vapour. The disadvantage of closed-cell foams is that they are more dense, thus requiring more material and thereby more expensive. The choice of foam should be based generally on the degree of insulation needed, environment, strength required, vapour control, available space, durability and so on. Extra costs can always be justified when compared to the need of a good job.

Both types of foams are commonly used in consumer, building and industrial applications and the choices are obvious. For example, you typically would not use open-cell foams where they are likely to absorb water, which would negate its thermal performance. On the other hand, they would be preferred to closed-cell rigid foams for packaging or for comfort applications. For each specific job, current technologies have provided many PUR systems in multiple, two-component, and three-component or even as single systems for coating large areas. For multi-component raw materials, it is very unlikely that all requirements can be purchased from one source as different manufacturers/suppliers will specialise in individual items and this system will be more cost-effective for foam producers.

7.3 Mixing ratio

7.3.1 Component A:B

Two-component PUR systems are supplied as component A (isocyanate) and component B (polyol blend) but will need a *mixing ratio* which will be provided by the supplier. It would be best if a foam producer considers this as a recommendation only and fine-tune in, in-house testing as the working temperatures and type of mixing/dispensing machines may be different. True, the supplier will provide information about specifications such as mixed temperature, viscosity, density and others in the certificate of analysis but if the materials are not used immediately, settling of chemicals, especially in component B, can take place. That is why pre-production mixing in the drums itself is recommended.

In purchases of two-component systems, the supplier will indicate specific mix ratios and speeds plus other relevant data like gel time, cream time, free-rise time and others. However, it is useful and important for a foam producer or technologists to know the calculations for arriving at the correct mix ratios. For this exercise the following data given in a certificate can be used:

System: isocyanate MDI (A) and polyol (B)
- Hydroxyl number of the polyol (component B)
- Amount of water present in this polyol as a percentage
- Isocyanate content (NCO group content) of MDI (component A)

Diphenyl methane isocyanate (MDI), nitrogen-carbon-oxygen (NCO) group, hydroxyl number (OH)

Formula: Amount of MDI = (polyol OH × 100) + (% water × 6,233) ÷ (NCO content × 13.35)

Example:
A polyol blend has an OH number of 98 and water content of 0.48%. The MDI used has a NCO content of 25%. What is the mixing ratio?

$$\text{Amount of MDI} = (98 \times 100) + (0.48 \times 6{,}233) \div (25 \times 13.35) = 38.3$$

Therefore, 38.3 parts of MDI are needed for every 100 parts of polyol.

7.3.2 Isocyanate index

A system supplier will also give an *isocyanate index*. This is a very important factor for using either toluene diisocyanate (TDI) or MDI to fine-tune a foaming process. This is where a pre-production 'cup-test' or a 'box-test' becomes very useful.

For example, if a system supplier gives an index factor of 98%, then the manufacturer should use 98% of 38.3 = 37.53 parts for every 100 parts of blended polyol. If the index is 103, then a foam manufacturer should use 103% of 38.3 = 39.45 parts for every 100 parts of blended polyol. Sometimes the index is given as 0.98 or 1.03, which means the same thing and that it has already been divided by 100.

7.3.3 Pump rates versus flow rates

Other than manual processing, all foam dispensing machines will have multiple tanks containing the different components connected to a central mix head via metering pumps and flow meters. Here, it is important to calibrate, set and carry out frequent maintenance checks to ensure correct volumes of chemical free-flow. With two-component systems, the problems are less as only two main tanks are involved.

Example:
From calibration and trials, it was found that a particular dispensing machine can metre 1,000 g/min of component B at a speed of 100 rpm. For a given formula, what is the speed in rpm at which it must metre to pump 2,100 g/min to make a foam batch?

Use the following formula: Known rpm ÷ known flow rate = rpm desired ÷ flow rate desired

$$100 \div 1,000 = \text{rpm desired} \div 2,100$$

Therefore, the pumping speed for B = 210 rpm.

Note: Use these data as a guide only, because pump flow rates are not always straight-line functions. Metering by set volume or weight is also an alternative.

7.3.4 Material flow versus mould volume

PUR two-component systems are used widely for moulded parts, especially for auto parts, footwear, sports goods and many others. It is useful to know the mould-filling basics. Consider a moulded product made with a mould having a volume of 3 cu.ft. and the mixing ratio of components A: B = 100:160. The density of the product made was 2 lbs per cu.ft. The processing data are as follows: Cream time = 30 seconds and Mould filling time = 15 seconds

Exercise: To determine
-   Total weight of components required:
    $M = V \times D, M = 3 \times 2 = 6$ lbs.
    A = 100/260 × 6 = 2.31 lbs. or 1.05 kg
    B = 160/260 × 6 = 3.69 lbs. or 1.68 kg
-   Throughput of each component
    Required machine output in lbs/min = 15 s × 6 lbs. = 24 lbs/min
    Component A – pounds per minute

    Use formula: Component A ÷ Component A + Component B = throughput A in lbs/min ÷ 24

Therefore, throughput of A = 9.23 lbs/min or 4.20 kg/min

Throughput of B = 14.77 lbs/min or 6.71 kg/min

Note: Depending on the product/process, a foam producer may want to make small allowances for material loss due to gas escape and material waste.

7.4 Compressed air

As we know compressed air is a valuable commodity in any manufacturing operation, more so for foaming processes. PU foam dispensing machines may be self-contained, if not then an air supply may be needed as well for air tools and other equipment on a factory floor. A small manufacturer may be able to manage with a single air compressor but it is more likely that a larger air supply system will be required. This can mean that an 'air compressor room', where several air compressors are connected to a large accumulator tank from which several supply feed lines will be available with suitable control valves.

General industrial ratio for compressed air is 5: 1, meaning a 5 HP compressor will yield about 1 HP of compressed air for industrial purposes. Compressed air generates heat and water concentration with perhaps other contaminates and these have to be removed for efficient dry air supplies by using water traps, valves and other devices at strategic places on line. The two important aspects of air supply in a factory are (1) pressure and (2) volume. However efficient an air supply line is, it is bound to have a 'drop' from source to the point of delivery and if factor is around 7–8%, it is considered acceptable.

Example for use as guideline:

Existing maximum pressure:	102 psi	7.0 bar		
Required pressure:	141 psi	9.7 bar		
Existing compressor capacity:	573 cfm	270.5 l/s	16.2 cu m/m	972 cu m/h
Additional capacity required:	191.9 cfm	90.5 l/s	5.4 cu m/m	324 cu m/h
Total required capacity:	765 cfm	361.0 l/s	21.7 cu m/m	1302 cu m/h

Therefore, the required increase in air volume = 33.5%

7.5 Calculating water requirements

Water is a requirement for any industrial operation, more so for manufacturing PU products. Since we are dealing with two-component systems, water will not be needed for formulating as it is already blended into polyol. However, water will be required for other areas like drinking, washrooms, flushing machine mixing heads and for precautionary measures such as showers, eye wash stations, laboratory tests and so on.

When designing a water supply system, a foam producer may find that more than one supply system may be needed.

This is because for some applications where machines, steam boilers or recirculating systems are involved, the water will have to be treated to remove the corrosive calcium content or other. For drinking purposes the water quality has to be even better. For general purpose uses, one may use any standard supply from main lines or other.

For small operations, one may opt for a well and overhead tank, with or without automatic operations. For larger operations, for cost-effectiveness purposes, one may opt for the same system but with larger or more tanks connected together. For very large volumes of productions, it is best to take the water supply from the main lines. It should be noted that although a factory will carry sufficient suitable fire-extinguishers (chemical and electrical fires), a certain amount of water may be needed for other purposes, including mopping up or other.

Basic calculations for water storage tanks are as follows:

1. Square or rectangular tanks: $L \times W \times H$ = volume (cu ft.) × 7 47 US gallons
   Example: 4 ft. × 4 ft. × 4 ft. = 64 cu ft. × 7.47 = 478 US gallons
   $$= 1{,}812 \text{ litres}$$
2. Cylindrical tank: $\pi \times r \times r \times h$ where, π = 3.1417 r (radius) = 2 ft. h (height.) = 8 ft.
   Example: 3.1417 × 4 × 8 = 100.53 cu ft. × 7.47 = 751 US gallons
   $$= 2{,}846 \text{ litres}$$

Note: Even when water is taken off a main line, it will have to be stored in tanks for easy gravity feed which can be created by larger diameter feed lines narrowing down to smaller diameter pipes. It is advisable to have an open/close tap at the bottom of each tank for periodic flushing out of the 'silt'.

7.6 Calculating electrical power

The principle of an electro-power system in simple terms is the generating station, the transmission lines, the substations and the distribution network. The generators produce high voltage electrical power, the overhead transmission lines move it to the regions to main stations and to substations, where the high voltages are 'stepped-down' and supplied to industrial, commercial and residential purposes. Some relevant keywords are amperes, voltages, current, single phase, three phase and 'step-down' transformers. A three-phase circuit is a combination of three separate single-phase circuits but they are conventionally connected.

Electrical power is probably the most important utility in any manufacturing operation where machinery is used. The need for use of single-phase or three-phase power depends on the type of power machines and equipment required to work efficiently. From a three-phase power installation, a single phase-power can be used, which is cheaper for working costs. However, the installation of three-phase supply will initially cost much more. In areas where power failures or short interruptions are frequent, one may opt to have a stand-by generator, even if, to run a few important machines. True, the subject of electrical power is a little complicated subject for most but for industrialists, small or big, it is useful to understand even the basics. The choice of using single-phase or three-phase power really depends on the types of machinery and other needs. Although, the installation of single-phase power is easier and cheaper, it is cheaper cost-wise to operate on a three-phase power. In electrical power, the amperage and voltages play the key role in supplying electrical power to work the motors, which are normally designated as –horse power (HP). While smaller motors can work on single phase, the larger ones will require three-phase power. For industrial purposes, the use of three-phase power is best. General internal factory wiring systems will be either 5 amp, 15 amp or 30 amp or other and probably it is best to have 30 amp and draw 5 amp or 15 amp. For long and efficient trouble-free power drawings, the correct gauges of wires must be used to prevent possible electrical fires.

Some machinery may carry more than one motor but the machinery manufacturer will specify the motors as single-phase 1 HP or three-phase 5 HP. In a manufacturing facility, the main items that will need electrical power will be lighting, motors and equipment. Voltages may range from 110/120 V or 230/240 V single phase and 440 V/480 V three phase. This will defer from country-to-country. Power is generally expressed as kilowatts (kW) and electrical consumption costs are based on a Unit = 1 kWH, which is the amount of electricity consumed in one hour.

A well-designed factory will have several power outlets at strategic places with easy access. The power board inside the factory must have a circuit-breaker as a protection against emergencies. For manual operations, where there are only small mixing machines, electric drills are used and for lighting, single phase should suffice. For all motors and equipment to be started at the same time, the installed power supply should be increased by at least 40–50% of the total estimated requirements. It is also prudent to have an extra capacity to meet future planned expansion, if any.

Useful data

1 HP = 746 watts   1 unit = kWH = kilo watts × one hour

Watts = volts × amperes × power factor

Ampere (l), when horse power (HP) is known:
- For single phase = (HP × 746) ÷ (E × Eff × pf)
- For three phase = (HP × 746) ÷ (1.73 × E × Eff × pf)

Where, l = amperes E = voltage in volts (120 V used in calculations)

Eff = efficiency as a decimal   pf = power factor as a decimal

If a motor is rated at a true HP, then it should deliver 746 watts of mechanical power. Single-phase motors are never 100% efficient in converting electrical energy into mechanical energy, so the amount of electrical power consumed by a motor is considerably higher than the mechanical power delivered. Because of losses from heat and friction means that a typical single-phase motor at best will be only 60–70% efficient. For small power tools operating with small induction motors, a figure closer to 60% efficiency, is more realistic. Hence, a genuine HP motor requires ≥ 1,250 watts of electrical power to deliver its rated power. In short, a 1HP motor will need ≥ 10 amps of current at 125 V or 5 amps at 250 V, to realistically deliver a true 1 HP from the same motor. Consider this as a general sound rule of thumb.

The power factor of an alternating current (AC) electrical power system is defined as the ratio of the active (true or real) power to the apparent power where (a) active power is measured in watts and the power drawn by the electrical resistance of a system doing useful work and (b) apparent power is measured in volt-amperes and is the voltage on an AC system, multiplied by all the current that flows in.

The following is a simple method of calculating power consumption:

Example:
A small foam manufacturing factory has
- One 1 HP motor
- One 3 HP motor
- 15 lights (each 100 watts)

Then, power consumption = $(1 \times 746) + (3 \times 746) + (15 \times 100)$ W
$$= 746 + 2{,}238 + 1{,}500 \text{ W}$$
$$= 4{,}484 \text{ W} = 4.484 \text{ kW}$$

Therefore, power costs will be based on approximately 4.5 units per hour.

Power consumption, if all work for 8 hours = $4.484 \times 8 = 35.872$ kW hours.
Now, W = volts × amperes × power factor

$$4{,}484 = 440 \times \text{amperes X } 0.8$$

Therefore, amperes = 12.74 amperes per phase
This factory can have a three-phase 440 V/50 Hz wiring system to suit 15 amp or 30 amp/phase (better) with a three-phase 440 V circuit-breaker trip switch with individual start/stop switches at point of power entry as a safety measure.

7.7 Formatting

PUs are made of a combination of many individual chemical components with polyols and isocyanates as a base and selected additives like catalysts, blowing agents, surfactants, fillers, stabilisers and others being used to complete a PU generating formula. By varying these components, it is possible to formulate to achieve flexible, semi-rigid, rigid, viscoelastic, microcellular and solid foams as desired.

The science behind this is polymer chemistry, which is little complex but for foam manufacturers at least a basic knowledge and formulating techniques are essential. Large volume foam producers, in all probability, will have an in-house polymer chemist, which is a big advantage but smaller foam producers may not be able to afford this luxury.

Since we are dealing with two-component systems, 'ready-made' to foam producer's specifications or supplier's standards, it is much easier for them. General practice is that component A will contain either TDI, MDI or a blend (isocyanates) and component B will have a fully blended polyol/polyols, meaning all the necessary additives have been added. With or without colour is a foam producer's choice. If not coloured, the resulting foam will be white.

These systems are very easy to process as the systems supplier will provide all relevant data such as viscosities, mix ratios, gel time, cream time, free-rise time, tack-free time, recommended temperature plus others. Most systems will indicate that densities can be varied by varying the proportions of A and B. It should be noted here that these variations can only be within reasonable limits. There are two other functions that a two-component user can do on a factory floor before processing. Add a desired colour pigment and increase the filler content by the addition of a biomass filler to increase IFD value by adding them to component B. This is where even a basic knowledge of the chemicals involved, their functions and the availability of in-house facilities of a cup-test or box-test becomes valuable to test different formatted formulae.

7.7.1 Basic components

Now, let us see what are the basic components needed for making different grades of PUs. Here, only the basic grades will be highlighted, which will suffice to give the reader a good understanding of the components needed for formatting. Each component is explained briefly just to identify only them and not their chemical structures or other.

7.7.1.1 Polyols

Most polyols are petro-based and flexible PUs are made from polyether polyols. Polyols are liquids and are generally available in drums or other packs. Different polyols with different functionalities give a range of varying properties and a polyol is the basic component on which a PU is formed. In a formula, whatever amount of polyol is used, for example, 80 kg or 50 kg, for formatting purposes it is taken as 100 parts by weight (pbw) and all other components are based on this factor.

Graft polyols contain copolymerised styrene and acrylonitrile. They are used in formulations between 10% and 45% to increase the load factor of a PU and may also be considered as filler polyols. Bio-polyols because of environmental concerns, non-petro–based polyols called bio-polyols are also being used by foam producers but they give lesser yields and also may have odour problems where masking bouquets may be required. In two-component systems, the fully blended polyol is classified as – component B.

7.7.1.2 Isocyanates

The two most used isocyanates are TDI and MDI, with the former being commonly used for flexible foams and the latter for moulded foams. They are highly toxic liquids and should be handled with care. In two-component systems, they are generally classified as component A. Sometimes suppliers of isocyanates will indicate an 'index factor' of 95 or 103 or other, which means when formulating one must use 95% or 103% of calculated value to obtain best results. We may take it as a factor for fine-tuning a formula to achieve quality foam. Also, higher contents of isocyanate in a formula will give lower densities.

7.7.1.3 Blowing agents

Water is normally used as the primary blowing agent in processing PUs for densities between 18 and 32 kg /cu.m. Water from normal municipal supplies would suffice, so long as there are no significant contaminants. If densities below 18 kg/cu.m is desired, a small percentage of methylene chloride can be used as a secondary blowing agent. As a general rule of thumb, the maximum amount that can be used in a formula is 5.0 pbw. of the polyol content. Some may advocate the use of higher contents of water but this will increase a risk of a fire hazard as the process is exothermic (heat giving). The water content generally decides the density range, with higher contents giving softer foams.

7.7.1.4 Catalysts

The production of PU foams requires at least two types of catalysts. This process involves two major reactions. In the polymerisation or the gelling reaction, the isocyanate reacts with the polyol to form foam. In the second, the gas-producing reaction or blowing reaction, the isocyanate reacts with the water to form polyuria and carbon dioxide gas. These reactions occur at different rates and will mainly depend on temperature and the type of catalysts used. For the first reaction, tin catalysts like stannous octoate are used, while for the gas-producing phase tertiary amines are used. These catalysts will control and balance both reactions and help to produce good cell structures.

7.7.1.5 Surfactants

Silicone-based surfactants are used in the manufacture or PU foams. Different grades are available for foam manufacturers to meet specific needs. Their main functions are as follows:
- Reduction of surface tension
- Resilience to prevent foam collapse, when foam is rising
- To control cell size for uniformity
- Silicone counteract the deforming effect of solids in the reacting system
- Surfactants will prevent bubble breakage at full rise of foam
- Stabilisation of cell walls

7.7.1.6 Fillers

Ideal fillers for PU foams are inorganic compounds of very fine particle size. These are added to PU formulations to increase density, load bearing and sound absorption. They also reduces costs and probably the most common one used is calcium carbonate. The author also suggests, from a particular point of view, that due to environmental issues, biomass fillers like rice hulls powder, wheat hulls powder and fine bamboo powder can also be used successfully. We must remember that although calcium carbonate is not petro-based, these biomass fillers mentioned are freely available and are a gift of mother nature and will not harm the Earth.

7.7.1.7 Pigments

If any colour is not included in the polyol content, the resulting foam will be white. Basically a yellow colour will protect the foam from UV action. Foam manufacturers use different coloured pigments to identify density, with common colours being blue, green and pink. Colouring of PU will not affect properties and will also add to its aesthetic value. Pigments selected should be compatible with the polyols used with high dispersion factor. If water-based, care must be taken to make a necessary adjustment to ensure not to exceed the maximum threshold limit. Some typical problems that may be encountered are foam instability, foam scorch, colour migration as well as abrasive action on pumps and mixers, if solid pigments are used.

7.7.1.8 Additives

Additives are materials that are added to foam formulations to achieve certain desired end-properties. These materials will not affect the general chemistry. Some of the common additives used by foam producers are as follows:
- Fire retardants
- Antioxidants
- Cell openers

- Plasticisers
- Antibacterial agents
- Anti-static agents
- UV stabilisers
- Foam hardeners
- Cross-linkers
- Compatibilisers

The best way to demonstrate formatting is to work out a practical example. Since flexible foams are the largest volumes produced in PUs, this presentation will be based on it.

Example:
A foam producer receives an order for cushions as follows
- Cushion size: 60 cm × 60 cm × 10 cm (thick)
- Quantity required: 100 nos.
- Density: between 27 and 28 kg/cu m
- IFD factor: between 2.2 and 2.30
- Should be flame retardant

Calculations: mass = volume × density Total mass = 0.036 cu.m × 28 kg/cu.m × 100
Total weight of chemicals = 100.80 kg add 1% (gas loss) + 3% waste factor
Therefore, estimated batch required = 104.83 kg

The foam producer refers to some existing formulae and works out a formulation. Because the customer's request is for flexible PU cushions, he decides on a combination of polyether polyol and graft polyol as a polyol blend for the starting base. He decides not to use calcium carbonate as a filler because of the presence of the graft polyol. All other components are standard ones and there is no need for the inclusion of methylene chloride as the density is on the higher range. From charts available, he works out the recommended amounts of water and TDI to achieve the desirable density. Since the foam should be fire retardant, a small quantity of an additive will be included.

So, the reasoning behind the final formulation is a polyol blend as the polymer base, water as the blowing agent, TDI as the isocyanate for the foam reaction, a surfactant to control the reactions and cell uniformity, two catalysts to promote and control reactions and an additive to make foam fire retardant. Since the customer requests high-quality flexible foam material, calcium carbonate as a filler is left out.

The final formulation to be used is as shown in Table 7.1 below.

To make sure of the end results, the producer carries out two or three box-tests using identical formula and once satisfied, prepares for production. First, the total weight of chemicals (104.83 kg) must be worked out in terms of the above-mentioned formulation, as shown in Table 7.2 below.

Table 7.1: Formula for production component.

Polyol	85 pbw
Graft polyol	15
TDI	40.70
Water	2.90
Surfactant	1.20
Amine catalyst	0.12
Tin catalyst	0.04
Methylene chloride	–
Flame retardant	3.00
Density (kg/cu.m)	27.20 kg/cu.m
IFD factor	2.30

Table 7.2: Raw material components calculation.

Component	Parts by weight	Proportion	Weight (kg)
Polyol	85.00	85/147.96 × 104.83	60.22
Graft polyol	15.00	15/147.96 × 104.83	10.63
TDI	40.70	40.70/147.96 × 104.83	28.84
Water	2.90	2.90/147.96 × 104.83	2.05
Surfactant	1.20	–	0.85
Amine catalyst	0.12	0.12/147.96 × 104.83	0.09
Tin catalyst	0.04	0.04/147.96 × 104.83	0.03
Flame retardant	3.00	3/147.96 × 104.83	2.12
Total	147.96	–	104.83

Note: Table 7.2 Example shown by author.

7.7.1.9 Viscosity

Viscosity is the resistance of a material to flow and is expressed in Pascal/second but is commonly stated as dyne second/cm which is called Poise – 1 Pascal = 10 Poise. When dealing with PU liquids, some of the factors that affect viscosity are chemical type, shear rate, temperature and pressure. Diluents may be used to increase flow but PUs needing good flow and exact ratio combinations to form good-quality foam has to be carefully formulated. In the case of mixing and pouring operations, viscosity may not be that important but when machines are used for dispensing and injecting; for example, viscosity of the two components will play an important role as pumps, flow meters and mixing heads are involved.

In general terms, the isocyanates are less viscous than the polyols and two-component PU system suppliers will always indicate the individual viscosities of each item and measured at a certain temperature. As we know, temperature can increase or decrease viscosities of liquids and that is why the foam structures of foams made at different room temperatures can differ in quality. When using machines for processing, it is important to maintain the metering pumps, flow lines and the mixing head in good stead at all times.

7.8 Conventional formulations

Formulations, different grades and so on for two-component systems have been presented in detail in the earlier chapters. Since we are dealing with processing of two-component systems, knowledge about conventional formulations may not be that important. However, for the benefit of the readers, a few conventional formulations based on actual practical manufactures are presented in Table 7.3 below.

Table 7.3: Basic formulations for flexible PU foams.

Raw Materials	kg				
Polyol (MW 3,000)	100	100	100	100	100
TDI (80/20)	55	50	47	42	38
Methylene chloride	6	5	3.6	–	–
Water	3.6	3.2	2.98	2.57	2.2
Surfactant (silicone)	1.2	1.0	0.92	0.82	0.72
Catalyst (tin)	0.25	0.23	0.21	0.20	0.18
Catalyst (amine)	0.23	0.24	0.24	0.27	0.27
Colour	0.38	0.35	0.30	0.28	0.25
Processing temperature 22–25 °C (72–77 °F)	–	–	–	–	–
Density (kg/cu.m)	18	21	23	28	32

Note: These formulations are given in good faith and no responsibility is taken by the author to say that they will work under different conditions. If necessary, a prior lab test like a cup-test or a box-test will help to fine-tune.

From these formulations, it can be observed that higher the isocyanate content, lower the density. Lower water levels will give higher densities. In order to lower foam costs, depending on the end applications, 10–40% filler (generally calcium carbonate) of polyol content can be added. Filler content can be much higher for cheaper foams. It is important to use pigments compatible with the grades of polyols used and should be mixed into it for best results. It has been observed that foams processed in cooler temperatures, for example, in mornings, give better-quality foams (Refer to the case in point presented under Chapter 10). Large volume foam producers may maintain cooler temperatures in the foaming area by air-conditioning or other means.

7.8.1 Foam cost calculation

The author would like to present a simple template in Table 7.4 for calculating foam costs.

Table 7.4: Foam cost calculation template – example 21 density.

Raw material	Kg (a)	Cost/kg (b)	Total cost (a × b)
Polyol	100	–	–
TDI	50	–	–
Methylene chloride	5.0	–	–
Water	3.2	–	–
Surfactant	1.0	–	–
Tin catalyst	0.23	–	–
Amine catalyst	0.24	–	–
Colour	0.35	–	–
Total weight	160.02	–	X
Weight loss (gas) 3%	4.80	–	–
Net weight	155.22	–	–
Less skins 10%	15.52	–	–
Good foam weight	139.70	–	–
Cost/kg	–	–	X ÷139.70

Note: Table 7.4 Example as compiled by author.

The above is a basic template. Material losses due to gas and skin losses (block trims) will vary depending on the processing parameters and methodologies used. This template can be applied to foam blocks made with two-component systems. The skins can be recycled and re-bonded into sheets for carpet underlay or thick slabs as mattress bases as a 'kickback' towards income/profits. For re-bonding, a foam producer can use either a steaming/compressing method or adhesive/compressing method as an economical recycling process.

7.9 Financial calculations

From these preceding chapters, readers would have gained a good knowledge of the technical aspects of PUs and the three chapters that follow will impart detailed knowledge of processing methods for two-component systems. Now, the author wishes to impart some knowledge on the 'business' side of manufacturing – *Marketing*. This information is especially useful for small foam producers or an entrepreneur, who may be able to process and manufacture good-quality products but still, they have to be sold at a profit for the business to be a success. It is hoped that the information provided here, based on both standard and practical 'hands-on' experience will be helpful.

Figure 7.3 shows some of the main forces that will interact with a business:

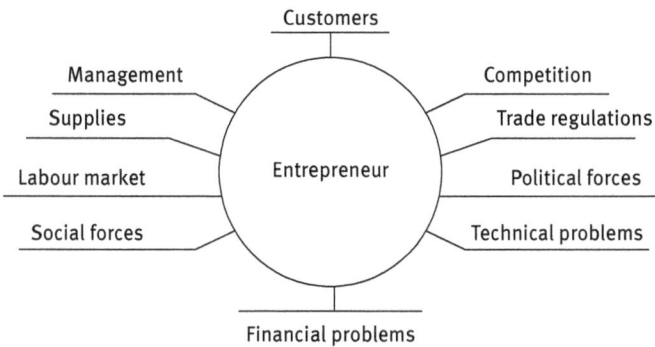

Figure 7.3: Some of the external forces influencing a business. (Compiled by the author).

As we know, there are many types of businesses and manufacturing PU foam is both exciting and challenging. The latter is due to the inherent 12–15% waste factor and also the need for safety as most of the chemical components used being toxic with possible fire hazards. The importance of this material on a daily need basis for all people and industries, negates such drawbacks and with proper handling and safety wear, these can be easily overcome, while the excessive waste factor can be reduced to reasonable levels by recycling and re-use.

Sound marketing practices are a key for a successful business operation. For a foam manufacturer, big or small, there are certain financial performance indicators like breakeven point, gross profit margin (GPM), return per kg, contribution margins and so on that will help monitor and enable action where necessary to keep the operation on a profitable level. Some of the important ones which influence a business are presented in Figure 7.4.

- Breakeven point
- Contribution margins
- Gross profit margins
- Cash flow
- Quick performance indicators

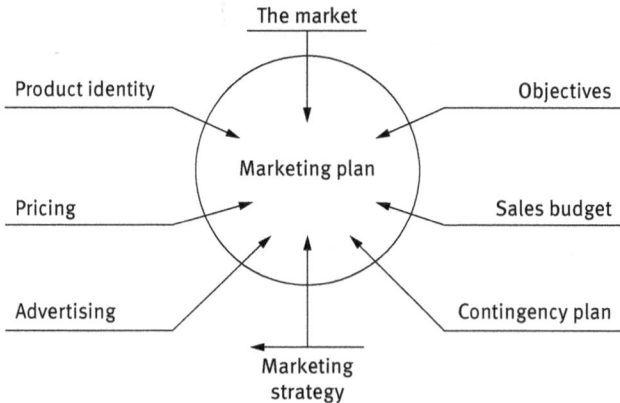

Figure 7.4: The influencing forces on a marketing plan. (Compiled by the author).

7.9.1 Breakeven point

This is a very important aspect to determine the feasibility of a project. This factor determines the total sales required to cover all overheads including both variable and fixed. This factor as a percentage will also indicate the minimum share of an intended or targeted market. BE is also identified as *breakeven analysis* and can be calculated graphically or using a standard formula as follows:

Formula: $BE = TFC \div (USP - UVC)$
where BE = breakeven
            TFC = total fixed costs
            USP = unit selling price
            UVC = unit variable cost

Example 1:
A foam manufacturer's total fixed costs for producing 50,000 foam sheets is $ 15,000 per year. The variable cost per unit is $3.50 made up of $1.10 for material, $2.00 for labour and $0.40 for overheads. The unit selling price of a foam sheet is $6.00.

Then, applying the formula – BE = 15,000 ÷ (6.00 – 3.50) = 6,000 units.

Therefore, the foam manufacturer will have to sell 6,000 sheets before starting to make a profit or he will make profits after a sales volume of $72,000.

Example 2:
A foam manufacturer makes foam cushions. Each cushion retails at $10.00. It costs $5.00 for each one and the fixed costs for the period is $750.00. What is the breakeven in units and in sales volume?

Applying formula, X units = FC ÷ (SP − VC)
= 750 ÷ (10 − 5) = 150 cushions = breakeven sales = $1,500

7.9.2 Contribution margins

A foam products manufacturer will normally be engaged in multiple-products activity such as making mattresses, cushions, sheets and so on. This will probably be true even for small producers. As the business proceeds, the owner through study and analysis may find some products making a strong contribution towards profits, whereas some may contribute much less. To maximise profits he decides to eliminate some of the products and uses the individual contribution factor.

A contribution margin is the amount of money a product generates towards meeting overheads and is calculated as the unit sale price minus the unit variable costs of a product. Some products may give high contribution margins but have smaller sales volumes, while some products with lesser margins will have very large sales volumes. Therefore, in decision-making all these factors should be taken into account to work out the best combination.

7.9.3 Gross profit margin

The GPM projected for 3–5 years in a feasibility analysis will project the gross profit possibilities as planned and will enable an entrepreneur or others to make adjustments where needed to achieve desired levels of profitability. The GPM will project the amount of money available to make payments such as bank loans, bank interest and partner's drawings plus others and indicate the final net profit before taxes. The GP can be expressed as a percentage as shown below:

Gross profit = (sales − cost of goods sold) ÷ sales × 100

7.9.4 Cash flow

This is a very important area of activity for any type of business. In simple terms, cash flow is the activity of total cash available versus payments. In accounting language many terms are used in compiling a cash flow forecast in the first instance and then in actual practice such as cash inflow, applications, expenses, outflow, disbursements and other.

Cash inflow may be made up of partner's equity, investments, bank loans, bank facilities, cash sales, credit sales while some examples of expense payments are bank loan instalments and interest, rent, salaries, partner's drawings, depreciation and so on. The first step in building a good cash flow is to ensure sufficient starting capital to include a contingency factor. Credit sales to be minimised but for a new business this might be a problem. If dealing with machinery and equipment, due to depreciation, some of them may have to be replaced down the road and a cash flow must also provide for this contingency. The depreciation factor can be taken as anywhere between 10 and 20 years.

However good a cash flow forecast is shown in a business plan, in actual practice, there may be certain periods where cash 'injections' are required to keep a positive flow. Options are partner's contributions or from banks which are better options and for this a business should work closely with their bank or banks. Their advice and guidance will at all times be valuable and they will also be anxious as you are to operate on a profitable level.

A sound business plan will carry a cash-flow forecast for the first 12 months, first 3–5 years and in actual operation must be on a cash accrual basis, meaning cash must accumulate.

7.9.5 Quick performance indicators

Instead of waiting until the periodic financial statements are made available to the owner or management to study the ongoing performance of an operation, one would like to suggest two indicators that an owner can use for constant monitoring. These will enable an owner to monitor performance and take corrective action if necessary. For foam manufacturers, the suggested monitoring tools can be very useful.

7.9.5.1 Cost per kilo
The total gross cost of operation versus. cost of raw materials consumed during a given period of time.

Example:
Total costs = ex-factory + marketing + administration + other = $60,000
Total raw material consumed during this period = 1,200 kg
Therefore, cost per kg = (60,000 ÷ 1,200) = $ 50.00 per kg
Compare with pre-set targets.

7.9.5.2 Return per kilo
The return per kilo is the total gross value of sales plus saleable goods versus raw material consumed during this period of time.

Example:
1,200 mattresses produced and sold @ $110 each = $132,000

Total raw material consumed during this production = 1,200 kg
Therefore, the return per kg = (132,000 ÷1,200) = $110 per kg

Compare with pre-set goals.
Here, for this calculation raw material consumed can be taken as:

Number of mattresses × unit weight per mattress + estimated wastage factor (%) = X kg

But for accounting purposes, a more accurate figure is required and the actual raw material consumed on a production floor can be calculated as:

Opening stock + RM drawn for production = Total RM available

Then, total available less work-in-progress (semi-finished) + RM unused = actual RM consumed

Bibliography

1. Gallagher Knowledge Center – 'Polyurethane Load Bearing Capacity' – www. gallaghercorp.com
2. FOAM-TECH – article 'Urethane foam-open cell vs. closed cell' – www.foam-tech.com/products/urethane
3. Defonseka, C.– 'Practical Guide to Flexible Polyurethane Foams' – 2013 Smithers Rapra Shropshire UK
4. Defonseka, C. – 'Practical Guide to Water-Blown Cellular Polymers' – 2016 Smithers Rapra Shropshire UK

8 Processing, moulding methods and troubleshooting

Two-component polyurethane (PU) materials such as polyols and isocyanates continue to grow in formula variations and these systems are expanding their uses with product manufacturers because of their improving performances over other materials. One current formula growth is in the development by many chemical manufacturers of 'green', eco-friendly and bio-based sustainable PUs as an alternative to petrochemical-based products. In many cases, these new formulations may affect the design of the metre/mixing/dispensing equipment possibly because of different flow properties.

The two-component PUs when processed on dispensing machines have characteristics that can cause metre/dispensing systems to differ from dispensing machines for other materials. The two-part PU equipment usually need to include different supply units, temperature controls, degassing equipment, air dryers and unique mixers. In addition, the processing of reactive PUs typically requires precision mixing and dispensing systems. Tight ratio tolerances and precise dispense volumes are often more important with fast-cure urethanes and high-speed automation applications. Suppliers of two-component PU systems may indicate in their specification sheets that it can be hand-mixed using a high-speed drill or mixer but they will also recommend the use of a dispensing machine for best results.

As we know, the range of PUs is vast from flexible to semi-rigid to rigid. There are many different ways of processing these PUs, for example, mix and pour, dispensing, moulding, spraying, casting, intermittent production where foam block are made one by one or slabstock production, where the foaming is continuous. To facilitate these processes there are standard methods as follows:

- **Manual Mix and Pour**: Hand-held or other mix and pour methods. For small-size products and small volumes, standard electric drills with suitable attachment would suffice.
- **Spraying systems:** These can range from small spray cans to on-site portable foaming systems to truck-housed systems for very large industrial applications.
- **Casting:** Self-skinning, rigid and flexible PU casting foams
- **Intermittent production:** Large foam blocks are made by semi-auto operated machines.
- **Dispensing:** Where the foam mix is metered through a mix-head. Here, there are three types of machine systems – low pressure, high pressure and automatic.
- **Moulding:** Moulding of auto parts, reaction injection moulding (RIM) and integral skin moulding.

https://doi.org/10.1515/9783110643169-008

– **Slabstock production:** Fully automatic continuous foaming on a moving conveyor, where multiple raw material components are metred through a mix-head
and poured directly on to the conveyor.

Since this book is dealing with two-component systems, slabstock production will
not be discussed. This chapter will present six processing methods as examples of
the above.

8.1 Manual mix and pour

MAXFLEX 422 is a fully formulated two-component PU system for making flexible
foam. It is ideal for production of furniture foam products, using open mould pour
technique.

8.1.1 Component data

Component A – (Isocyanate 117) Component B (Maxflex 422)
Mix ratio   55   100
Viscosity @77 °F   210 cps   400–600 cps
Cream time   22–28 s
Gel time   123–133 s
End of rise   143–153 s
De-moulding time   7–10 min.
Free-rise density   76.8–83.2 kg/cu.m (4.8–5.2 lbs/cu ft.)
Moulded density   83.2–90.6 kg/cu.m (5.2–5.6 lbs/cu ft.)
Processing temperature around 72 °F

8.1.2 Product

Foam block size: 30 in. L (75 cm) × 30 in. W (75 cm) × 40 in. H (100 cm) to cut cushions for the manufacture of furniture. Foam block size can vary according to cushion size requirements but weight, shot size and mixing cylinder volumes would be
factors.

8.1.3 Moulds

Made of wood 3/4 in. (2 cm) thick with all sides detachable. Bottom plate to have a
3/4 in. depth groove to accommodate vertical sides when assembled. Mould to have

a clamping arrangement to prevent leaking of liquid mix. A suitable lightweight 'floating lid' size 29 in. (72.5 cm) × 29 in.(72.5 cm) × 1/8 in. (0.31 cm) thickness will be needed to negate the meniscus being formed during the rising of foam. The number of moulds will depend on the volume of production but it is advisable to work with at least two moulds. If moulds are mounted on wheels, it would help for easy movement. The insides of the mould should be very smooth (laminated wood, lined with aluminium sheet or other) to minimise the material loss due to adhesion to sides. To compensate for material adhesion to sides and foam shrinkage, generally 0.5 in.(1.25 cm) thick trims would be standard but this could be less. A pre-production box-test will confirm the actual trim size needed.

8.1.4 Material requirement calculation

Volume = 31 in. × 31 in. × 41 in. = 22.80 cu.ft
Apply, $M = V \times D$ M = weight V = volume D = density
Then, weight = 22.80 cu.ft. × 5.2 lbs/cu.ft. = 118.56 lbs = 53.89 kg
Add, 3–5% to compensate for gas loss and waste = 53.89 kg + 2.69 kg = 56.58 kg
Now, mixing ratio – A:B = 55:100
Therefore, A = 55 ÷ 155 × 56.58 = 20.08 kg
            B = 100 ÷ 155 × 56.58 = 36.50 kg

8.1.5 Processing method

Weigh components A and B accurately into two separate vessels. Add a very small amount of colour (yellow or other) to component B and mix well for a short time. Spray a release agent like silicone or other to the insides of the assembled mould. Add component A to component B and mix for 2–3 seconds and immediately pour the mix, still in the liquid state, into the open mould. If there is a delay in pouring, the mix will start to 'cream' and the whole batch will be lost.

   The liquid mix at the bottom of the mould will turn a 'cream' colour after few seconds and will start to rise slowly. When the foam mass has come up about one-third of the height of the mould, place the 'floating lid' on top of the rising foam to prevent a meniscus (rounded top) being formed. When the foam block surface is not tacky, de-moulding can take place. From a holding area, remove these semi-cured blocks to a well-ventilated room/area and place them about a foot apart. Do not stack one on top of another to prevent fire hazards due to the exothermic (heat giving) reactions still taking place inside the foam. After 24 hours, these foam blocks can be taken for fabrication. When blocks are being stored, it is recommended that the first in–first out (FIFO) system be used to ensure that only fully cured foam blocks are taken.

8.1.6 Cutting and fabrication

All sides of a foam block has to be trimmed and this can be done by standard band saw machines with hot-wire cutting an option, if the local regulations allow it. These same machines can cut the foam blocks into any sizes and shapes desired.

8.1.7 Recycling of foam waste

Foam wastes can be shredded into small pieces using shredding machines and then re-bonded into blocks using either an adhesive/compression or steam/compression systems or other systems. First, they are compressed into large blocks and then cut into sheets for carpet underlay or into thick slabs as mattress bases. This would also enhance income contribution towards profitability.

8.2 Spraying systems

Spray polyurethane foam (SPF) is made by two chemicals mixing and reacting to create a foam. In this case, the two chemicals being a polyol and an isocyanate will react very quickly expanding on contact to create suitable foams to insulates, air seals and also provides a moisture barrier. SPF insulation is known to resist heat transfer extremely well and also offers a highly effective solution in reducing unwanted air infiltration through cracks, seams and joints. There are different types of SPF foams but here, we will discuss the most common ones used by professional contractors being either high-pressure or low-pressure foams.

Whether retro-fitting a home or building a new one, many prefer SPF foam insulation as a great way to save on energy costs and improve comfort. One has a choice of 'open-cell' or 'closed-cell' foam. There are several major differences between the two types, leading to advantages and disadvantages for both depending on the end application. A comparison of both is shown in Table 8.1.

Table 8.1: Insulant properties of closed/open cell spay foam.

Closed cell	Open cell
Higher R-value (greater than 6.0 per inch)	R-value (approx. 3.5 per inch)
Lower moisture vapour permeability	Higher moisture vapour permeability
Air barrier	Air barrier at full wall thickness
Higher strength and rigidity	Lower strength and rigidity
Resists water	Not good for applications in direct contact with water
Medium density (1.75–2.25 lbs/cu.ft.)	Low density (0.4–1.2 lbs/cu.ft.)
Absorbs sound	Absorbs sound much better

In principle, two chemicals in liquid form combine to create a foam. Instead of the standard TDI–polyol combination for spray foams the combination would generally be a MDI–polyol combination, which are called the 'A' side and the 'B' side. The common PU spray systems will consist of methylene diphenyl diisocyanate (MDI) and polymeric methylene diphenyl diisocyanate (pMDI) as the A side. The B side will typically be a blend of polyols, catalysts, blowing agent, flame retardant and surfactant. The polyols will form the basic polymer matrix, while the additives will have their own individual functions to create a stable and efficient insulating foam.

8.2.1 Application methods

Spray foam systems can be categorised into open-cell spray foams (ocSPF) and closed-cell spray foam (ccSPF). Spray foam insulations are applied using two main delivery systems, low-pressure kits/systems for small applications and high-pressure systems for industrial or large building applications using large drums for chemical supplies.

For small- to mid-size projects, there are low-pressure (typically less than 250 psi) two-component kits. These can be used to seal and insulate small- to mid-size areas around a home, for example, such as attics, crawl places and rim joists. These should be handled by professionals or at least by people who are familiar with these foams where safety wear will be a must.

Two-component, high-pressure systems (typically 800–1,600 psi) generally use 55 gallon drums for the chemical supplies and used on large areas of construction or major renovations on walls and roofs. For these applications specialised training is required for personnel, and safety wear is a must with respirators and goggles. These systems will basically consist of a spray rig (mini-truck) which will house the chemical supplies, air supply and other items such as delivery hose, up to about 300 ft. (93 m) and adjustable spraying nozzles with long handles. Smaller portable systems are also available, depending on the end applications.

Although, the chemical mixture when applied flows freely and hardens quickly, the fumes from the application will linger around the area for some time. Since inhalation is not good, people should adhere to the advice of a contractor or professional who carries out the job.

8.2.2 Two-component polyurethane system for insulating large areas

BAUMERK Construction Chemicals supplies a special two-component PU injection resin system for application in very large areas. PUR IN 24 is a PU-based two-component injection system which reacts with water quickly and increases a system's volume by 10–15 times.

8.2.2.1 Fields of applications
- To stop water leaks from cavities and cracks of concrete walls and floors of tunnels, bridges and similar constructions
- Insulation of cold joints
- On water tanks and swimming pools
- Insulation of interior basements

8.2.2.2 Features and benefits
- Stops water leakage on applied surface and provides water isolation
- Fills the holes of system without losing volume
- Can be used in moist concrete
- Blocks negative water flow

8.2.2.3 Technical data
Table 8.2 shows Technical Specifications of Product

Table 8.2: The basic technical data of this system.

Content	Polyurethane-based
Colour	Brown/yellowish
Density	1.1–1.16 g/cu. cm
Mixing	Comp. A 10 kg: Comp. B 1 kg
Flash point	>100 °C (212 °F)
Application temperature	+ 10 to + 30 °C (50–86 °F)
Drying time	Touch dry= 20–30 minutes, Full cure 7 days
Foam expansion time	15 seconds
HS Code	3208.90.91.00.23
Application method	Mechanical pump
Shelf-life	12 months (unopened containers)

8.2.2.4 Application method
For best results, prior to injection of material, the cracks to be worked on are applied with EPOX repairing mortars. Along the course of the cracks holes are drilled at a distance of 10–15 cm to each other on alternating sides of the crack. Packers are inserted into the injection holes and preferably injection to be done from bottom to the top.

Component A and B are mixed separately for about 30 seconds at low speeds. Then component B is added to component A slowly and mixed together for about 3–4 minutes till a homogeneous mixture is obtained. PUR 24 is then injected using a mechanical pump under low pressure. When this low viscosity liquid mixture is applied, it expands rapidly and it becomes a solid foam in about, 4–5 seconds.

8.3 Self-skinning, rigid and flexible polyurethane casting foams

Polytek Development Corp. (USA) offers an exciting and versatile range of two-component systems under their brand name *Polyfoam Series* for making self-skinning, rigid and flexible casting foams. The Polyfoam Series consists of parts A and B in liquid form which can be used to cast rigid or flexible products with densities in the range of 3–20 lbs/cu. ft. (48–320 kg/cu m). Polyfoams can be ideally used for making decorative objects, lightweight mould shells, production parts, models, patterns, fixtures and general tooling use. Polyfoam systems are practically odourless and do not contain toluene diisocyanate (TDI), heavy metals or HCFCs. Polyfoam R-2, R-5 and R-8 are rigid foams, while Polyfoam F-3 and F-5 are flexible for casting soft parts.

8.3.1 Mould preparation

Polyfoams will reproduce minute details from moulds or patterns but may stick when poured onto improperly prepared surfaces. Spraying a release agent prior to pour is an option. There are many materials from which suitable moulds can be made and moulds made from polyethylene and silicone rubber made from PlatSil 71 or 73 series will not require a release agent. A suitable barrier coat such as Barrier PF will help to extend a mould life when using rigid foams. For best results, a release agent should be applied to the surface before applying Barrier PF. Metal moulds or PU rubber (Poly 74 or 75 series), for example, should be dry and coated with a suitable release agent such as paste wax, polycoat or PVA solution. If rubber moulds are used, they must be hard enough so as not to distort under moulding pressures.

8.3.2 Polyfoam compaction calculation

Calculate the volume of the mould space to be filled with foam in cubic inches. Determine the density desired of foam part in pounds per cubic feet. Divide the desired density by 1,728 cubic inches. The result will be a decimal 'factor'. Now, multiply the volume of the space to be filled by this factor. This will give the amount of pounds of Polyfoam liquid needed. You may compensate for gas loss and material waste (if any) as an option. To convert lbs/cu. ft. to kg/cu. m multiply by 16.02.

8.3.3 Mixing components

These resins sets fast and it is advisable for all materials, moulds, equipment or others that are made ready for immediate processing. Make sure that both parts A and B are at room temperature and weigh both parts separately and put them into polyethylene pails (ideal). Combine both parts and mix quickly using a turbo mixer or other high-speed mixer for 15 seconds. Pour mix into the mould cavity as quickly as possible in a liquid state. If there is delay, foaming will start in the container itself and the full batch will be lost. Table 8.3 shows the physical properties of the polyfoam range:

Table 8.3: Polyfoam physical properties Reference Technical Bulletin – Polytek Development Corp.

Polyfoam Physical Properties

Grades	R-2	R-5/R-8	F-3	F-5
Mix ratio	1A:1B	1A:1B	1A:2B	1A:1B
Mix viscosity (cP)	500	1,100	2,000	1,400
Cream time (s)	30	45	25	45
Rise time (min)	3	2	1.5	3–5
Tack-free time (min)	10	3	3	25
De-mould time (min)	30	10–15	10	30–60
Free-rise density (kg/cu. m)	40	R5-80 R8-128	48	80
Moulded density (kg/cu m)	64–128	128–320	80–128	128-240The

Polyfoam physical properties-reference Technical Bulletin –Polytek Development Corp.

8.3.4 Curing procedure

Packing polyfoams to a minimum of 2–3 lbs/cu. ft. (32–48 kg/cu m) above their free-rise density is recommended by the system suppliers to achieve good mould-fill and thus minute details of product. A lid with small vents to allow gas to escape as foam rises, should be firmly clamped before the foam rises. Once the foam rises, any movement of the mould may tend to promote cell collapse. Castings should be allowed to be in the mould until fully cured as de-moulding too soon will result in product deformation. For best results, the mould should be warmed to 75 to 85 °F prior to casting the first part. Once heated, the mould will retain sufficient heat for continuous production.

8.3.5 Finishing products

Cured Polyfoam products will tend to become yellow and chalk when exposed to sunlight and should be painted or sealed for exterior use. Products made from Polyfoam systems R-2, R-5 and R-8 can be easily drilled, sanded and machined. If a casting is to be painted or coated, the adhesion factor should be carefully checked for suitability and durability. When casting rigid foams, the use of an appropriate primer/barrier coat such as Barrier PF, sprayed into the mould and allowed to dry before casting. This procedure will help in better adhesion when painting or coating.

8.3.6 Storage life

It is good for at least six months if stored at a temperature (60–90 °F) and the containers are unopened. Once containers are opened and used the balance must be resealed tightly as atmospheric moisture contamination may degrade the materials. Polyfoam F-5 may crystallise, develop sediment and become cloudy if stored at temperatures below 60 °F. To restore the product to its original state, loosen the lid slowly (to let-off any pressure built-up) and warm product to 120–160 °F until liquid is clear. For this, one may use a thermo-controlled jacket and before use, let the material cool down to room temperature.

8.4 Intermittent process

This process will produce large foam blocks but one at a time suitable for making flexible PU foams for comfort applications. Although, common practice has been to produce these large foam blocks on multiple chemical streams, where different densities and properties could be made by varying the formulations on the machine itself; when using two-component systems, only one density and formulated properties can be made per block as supplied. The density may be varied within reasonable limits by varying the mix ratio. The big advantage of a two-component system is zero mistakes in chemical weighing, ease of use, ease of storage and so on.

8.4.1 Foam blocks for mattresses

Once a large foam block is made, it can be converted into mattresses, cushions, sheets, wedges, slabs or other. Since there is an inherent waste factor in standard PU foam productions, where this waste could be reduced are good mould design, correct mix ratios, mixing times and efficient pour/dispensing practices. For blocks, most wastes results in formation of a rounded top (meniscus), which has to be eliminated. The waste losses due to block trimmings cannot be avoided but can be minimised. The use of under-cured foam blocks will also result in unnecessary wastes.

8.4.2 Polyurethane foam mattresses

Foam mattresses may be broadly categorised as luxury, high-end and low-end, the difference being in its construction. Probably the most common mattress when taken on a global scale would be the low-end single foam mattress with a cover. These foams can be in different densities with viscoelastic foams (memory foams) being used for luxury mattresses.

Let us discuss the production of a foam block to make standard foam mattresses.

Product: Standard foam mattress
Size: 39 in. (97.5 cm wide) × 75 in. (187.5 cm length) × 4 in. (10 cm thick)
Number of mattresses: 10 per foam block
PU System: Two-component. Specifications as shown below in text
Mould construction: As shown below in text
Machinery: As shown below in text
Processing method: As shown below in text
Fabrication: Machinery as shown below in text

8.4.2.1 Raw material system
The two-component PU foam system suitable for flexible foam blocks are used for making mattresses and furniture products. Table 8.4 shows the technical specifications as given by the supplier's technical data sheet.

Table 8.4: Technical data as given by Premilec Inc.

Component	A	B
Viscosity @77 F	210 cps	400–600 cps
Mixing ratio	55	100
Cream time (s)	22–28	–
Gel time (s)	123–133	–
End of rise (s)	143–157	–
De-mould time (min)	7–10	–
Moulded density (kg/cu. m)	76–83	–

These systems are delivered in 45 gallon sealed coloured drums, red for component A and either green or blue for component B. For best results, the contents of both drums should be mixed at slow speed for a short time before connecting to foaming machine.

8.4.2.2 Required foam block size

Based on ten mattresses per foam block, the foamed, untrimmed foam block will be:

Each size of mattress (trimmed) = 39 in. × 75 in. × 4 in.

Therefore making an allowance of 0.5 in. for skins on 5 sides + 1.0 in. for top, the size of an untrimmed foam block will be = 41 in. × 76 in. × (10 × 4 in. + 1.0 in.). It should be noted that when cutting with band saws, it is not possible to get 10 × 4 in. thick mattresses from 40 in. foam block height due to material loss because of the thickness of the band saw blade, hence the extra allowance.

Therefore, the actual size of an untrimmed foam block will be = 41 in. × 76 in. × 41.5 in.

8.4.2.3 Required amount of raw materials needed per block

On basis of component A and component B with mixing ratios: A:B = 55:100
Density = 83 kg/cu m (5.2 lbs/cu. ft.)
Applying formula $M = V \times D$ $M = 41 \times 76 \times 41.5$ in. × 5.2 lbs/cu. ft.
M (weight) = 74.83 cu. ft. × 5.2 = 389.1 lbs. = 176.9 kg
Add 1% for material loss due to gas = 178.67 kg
Component A = 55÷155 × 178.67 = 63.40 kg Component B = 100÷155 × 178.67 = 115.27

8.4.2.4 Mould construction

This can be made of wood, aluminium, laminated board or other materials with sufficient thickness to withstand the pressure build-up due to the exothermic (heat-giving) reactions of the chemical mixture. The mould should be rectangular, with all four sides detachable for easy removal of foam block. The inside walls should be smooth to minimise the material loss due to foam adhesion and the bottom board must have grooves to accommodate the vertical panels when assembled. All walls must lock tightly to prevent material seepage. The mould should be mounted on wheels for easy movement. When assembled, the inside volume of the mould should be 41in. wide × 76 in. length × 41.5 in. height or more. The number of moulds will depend on the volume of production but it is best to have at least two.

8.4.2.5 Foaming machine

The machine selected should have a 'shot' capacity or dispensing capacity of at least 190 kg to accommodate both components and be able to mix and dispense the mixture of 178.67 kg as a single volume. Figure 8.1 is a semi-automatic foaming machine suitable for this processing.

This is a very versatile and effiient machine capable of producing 12 foam blocks per hour and does not need any special skills to operate. The individual components

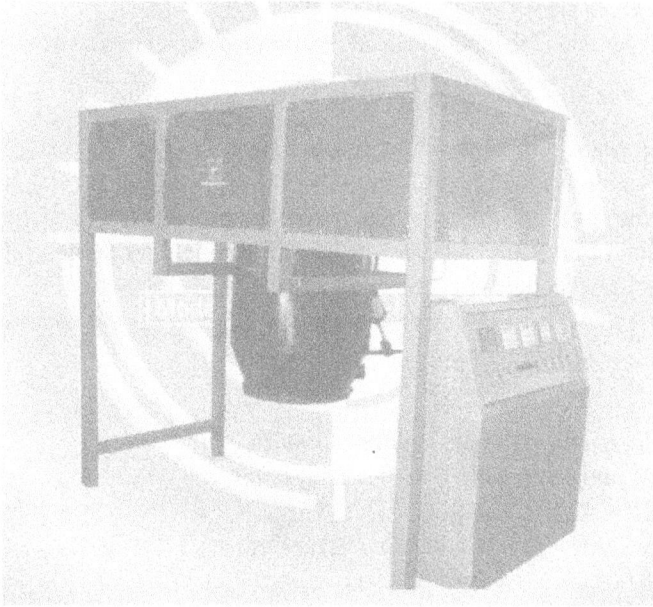

Figure 8.1: Single Foam Block Making Machine.
Source: Reproduced with permission from Modern Enterprises, India.

can be pre-set by weight and streamed into the large mixing chamber. All other important processing parameters can also be pre-set. This machine is actually designed for
multiple component production but can also be modified for two-component systems.

8.4.2.6 Processing method
Both components should be mixed at slow speed in their containers and then connected to the machine. Some machines may have large volume tanks for components
A and B, which can hold the volumes of several drums. This can greatly help in longer production runs. As an example, use the following processing parameters:

Component A – 63.40 kg Component – B 115.27 kg
Mixing speed: 3,000 rpm Mixing time: 4 seconds
Cream time: 25 seconds Gel time: 125 seconds
End of rise: 150 seconds De-mould time: 10 minutes

It is recommended that safety wear like goggles, respirator, gloves and an apron be
worn.

Set the weights for A and B and mixing speed. Bring the mould on wheels and
position it under the vertical mixing vessel of machine. Lower the vessel and position

it about 15–20 cm above the floor of the mould. Stream in component B into a mixing vessel and mix for 30 seconds. Set the mixing time for 4 seconds in auto mode. Stream in component A and activate mixer, which will mix for 4 seconds and discharge the full mixture onto the floor of the mould by auto-opening of the bottom valve. When fully dispensed, the mixing vessel will automatically rise vertically.

The liquid mixture on the floor of the mould will start to 'cream' and slowly rise going through the 'gelling' period. When the foam has risen about one-third of the height of the mould, place the 'floating lid' on top of the rising foam. Do not move the mould or do anything to disturb the rising foam nor inhale the fumes which will be toxic.

When the foam surface is not tacky, de-moulding can take place and the foam block should be kept in a temporary holding place before transferring to the main storage for final curing. After a minimum period of 24 hours these blocks can be taken for fabrication/cutting.

8.4.2.7 Fabrication/cutting
The fully cured blocks are first trimmed on all sides and the trimmed block should now have dimensions: 39 in.(97.5 cm) × 75 in.(187.5 cm) × 40 in.(100 cm) from which the ten mattresses can be cut. Figure 8.2 shows a band saw cutting machine

Figure 8.2: Vertical Cutting Machine.
Source: Reproduced with permission from Modern Enterprises, India.

that can be used. A foam producer may opt for two machines – (1) vertical cutter as shown and a (2) horizontal cutter for more efficient operation.

8.4.2.8 Finished products

Once the foam mattresses are cut and passed onto the finishing department, it is best to at least carry out basic quality control and property checks such as density, IFD (support factor), dimensions, grading if any, before suitable covers are put on and final packing. The IFD can be checked with a small hand-held device. These information will be valuable as marketing tools. It should be noted that if band saw cutting machines are used, due to the thicknesses of the blades, it may not be possible to cut 10–4 in. thick mattresses and a suitable adjustment must be made in the amount poured to extra height.

8.5 Dispensing two-component polyurethane systems

Here, we will discuss low-pressure and high-pressure dispensing as automatic dispensing is more or less an extension of them. The two-part metering technology selected for PU dispensing systems is the most critical component of the machine. The metering unit controls the flow of the polyol and isocyanate materials. Metering technology selection is based on several variables, including flow rate, volume dispensed per application, mix ratio by volume and accessory requirements. Selecting the right metering technology determines the performance of the entire dispensing systems.

Probably the most common metering equipment types for dispensing reactive PUs are defined by two categories of meters: (1) continuous flow meters and (2) shot dispensing meters. Each type has its own design for process control and matching application requirements based on the volume of material required per product to be manufactured.

Meters can be defined as either fixed or variable ratio, with each type having its own distinct advantages. Fixed-ratio meters will prevent changing or tampering with a given ratio and will need a mechanical component replacement to change the ratio. Variable-ratio meters with one-drive motor will require a mechanical adjustment to change the ratio. Variable-ratio meters with dual-independent drive motors (each with own speed controls) will use a simple electronic adjustment in the control panel.

The purpose of selecting variable-ratio meter machines is to be able to adjust the ratio for variations of a current formula, for example, to vary the foam density or when needing a new ratio requirement when changing material suppliers. The variable-ratio meters are excellent systems for formulators who want a lab unit for analysis of various two-component chemistries.

8.5.1 Low-pressure dispensing

There are many types of low-pressure (LP) dispensing machines from different man-ufacturers but the basic operating principle is more or less the same. Consider the following LP ECD-Basic Series Machine foam Edge-Sweets Company (ESCO) USA:

8.5.1.1 Low-pressure machine
ESCO's ECD basic machine series are good solutions when simple dynamic metering and dispensing solutions are required. The ECD machine features a control station mounted at the mixing head, within easy reach of an operator. This versatile machine offers dynamic mixing of two-component PU systems.

Machinery throughput specifications
- Ranges from 1 to 20 pounds (0.45–9.10 kg) per minute at a nominal material ratio of 1:1 (isocyanate: polyol)
- Ratios are variable within individual metered stream output capacities

Mixing head specifications
- ESCO series M200-2PE two-component dynamic mixing head
- Three-way ball valves for material injection and recirculation
- Diaphragm-style solvent injector valve
- Gear-style mixer with lip-style seal
- AC direct drive with 1,000–3,500 rpm variable speed range
- Return line valves for pressure balancing
- 5 gallon stainless steel tank 110 psi rated for solvent flush system

Products
Sports shoes, shoe soles

8.5.2 High-pressure dispensing

Hennecke PU technology offers a top of the league PU processing machines that have always been a synonym for top-class high-pressure PU processing and the heart of countless PU systems. The design of the TOPLINE series comprises sophisti-cated arrangement of the individual units, top-quality components, an ultra-mod-ern, user-friendly control system and state-of-the-art mix head technology.
Some of the main applications are as follows:
- Rigid foam insulation with densities from 20 to 200 kg/cu. m

- Rigid integral skin foam with massive border zone, cellular core and densities from 180 to 1,100 kg/cu m
- Semi- rigid integral skin foam with densities from 150 to 1,100 kg/cu. m
- Semi-rigid filling foam with densities from 100 to 300 kg/cu. m
- Cold-curing flexible foam with densities from 20 to 70 kg/cu. m
- Hot-curing flexible foam with densities from 25 to 50 kg/cu. m

8.5.3 End applications

Some of the products are pipe insulation, auto panels, steering wheels, car seats, foamed panels and others mostly for industrial, building construction and automotive applications.

8.6 Moulding with two-component systems

Processing equipment basically consist of a suitable dispensing machine and a mould or moulds for multiple-moulding. Moulds can be single cavity or multi-cavity, with or without inserts made of aluminium, steel or other material and in two halves. For good quality production the mould design will be very important, with provision for venting for the escape of gas/air, with good flow paths and clamping arrangements. The inside surface of the mould will be treated to yield a smooth, matt or other desired effect and can include patterns and paintable techniques also. Moulds will tend to have a 'workable' span, meaning they can be used, for example, 30,000, 40,000 or 50,000 mouldings or other before needing repairs or rejection.

In order to make good quality parts, it is important that the dispensing to be used should be calibrated before use. When dealing with two-component systems, it is so much easier for production but the basics must be attended to. First, the dispensing machine must have a greater 'shot' capacity than the amount of mixed material needed to fill a mould. Next, setting of the mixing ratios at a given temperature becomes important. Material supplier's data sheets can be used as a guide. If the actual production temperatures of the materials are different to the ones set, then the machine ratios must be reset again. If different two-component systems are to be used on the same machine, the mixing ratios must be reset to each system before processing.

After the machine has been calibrated, the last test before commencing production is a cup-shot or free-rise foam test. This test is done mainly to check whether the foam re-activities are normal. The simple cup-shot test can catch many of the common chemical and equipment problems which can lead to production rejects. A cup-shot test can be done by pouring a small shot from the machine mix head into a pre-weighed paper cup. The shot amount must be sufficient to expand and slightly overflow over the top of the cup. Four tests can be carried out with this test: *foam re-activities, throughput, free-rise density* and *foam structure.*

8.6.1 Foam re-activities

When the two components are streamed into the mix head of a dispensing machine, within seconds the resulting mixture must be dispensed into the mould, still in the liquid state. The chemical reaction must start inside the closed mould. The foaming process will undergo four stages as follows and observed during the cup-shot test:

- **Cream time:** It is the time in seconds from beginning of the pour until the mixture starts to expand. It will change to a creamy/milky colour and start to rise.
- **Rise time:** It is the time from beginning the pour until the foam stops rising. At this stage, the foam would have slightly overflowed and it will look like a 'muffin'.
- **Tack-free time:** It is the time between pour and cure, until the foam surface is no longer tacky.
- **Pull time:** It is the time between pour and cure, until the top of the foam can be pinched and pulled without tearing.

These reactivity stages are like a 'fingerprint' for two-component PU systems. Naturally, different systems will have different reactivity times and if the chemistry is correct and the machine is working well, the same results should occur for any number of pours of a production run. Since we are dealing with chemicals, processing temperatures become important as well as working within tolerances for finished products. From the actual parameters obtained from the cup test which can be used as guidelines, good production runs can be set up. In-mould product cooling time will be based on the tack-free time but will also depend on the mould design and size of products being moulded. Figure 8.3 shows the sequence of a cup test:

Figure 8.3: Sequence of cup test.
Source: Adopted from free technical Lit provided by BASF.

8.6.2 Calculation of throughput

The throughput or shot capacity becomes important when deciding how much material is to be injected into a mould. All machines will specify this data but may differ under different working conditions and temperatures. Here, the cup-shot test is helpful. Weigh the cup before and after foaming and then divide the total material by the flow time. This known as the shot time and is measured in grams per second.

8.6.3 Free-rise density

This will indicate the mould fill time and also the foaming power of a particular two-component system used. There will be a slight difference between free-rise density (hot) and the final density (cold) being more. This will differ with each system and can be compared to the data indicated in the specifications sheet supplied by the systems supplier.

Calculation:

- Cut off the top of the foam, so that it is flush with the top of the cup
- Weight of foam = total weight (cup + foam) – weight of cup
- Apply, $M = V \times D$, where D = free-rise density(g/cc)

8.6.4 Foam structure

Cut the foam mass lengthwise and in quarters. The foam should have a uniform cell structure and also have a smooth velvety texture. Swirls, splits or other imperfections are usually signs of machine or equipment problems but with reputable machinery and will be rare. For possible solutions to these problems refer to the troubleshooting guide shown later in this chapter. Commence production after the recommended solutions are done and are satisfactory.

8.6.5 Filling the moulds

The heart of the production process is the mix head of a dispensing machine. This needs good maintenance and some advanced machines will have auto-flushing. This is to avoid any leftover material mix foaming and complicating the next shot. There are two methods of filling moulds: (1) *open pour* and (2) *direct injection.*

- Open pour – The mixture is poured by the calculated volume into an open mould and the mould is closed. This can be done by hand or the machine can be programmed to release a calculated volume as a single shot.

–   Direct injection – The mixture is directly injected into a closed mould through a small opening/hole or an injection port.

8.6.6 Recommended pouring patterns

Before pouring starts, make sure all equipment is working properly and the mould temperatures are right, clamping and cooling systems and other processing aids are in good order. The main purpose of a good pouring pattern is to ensure that sufficient material volume is poured/injected to fully fill the mould as a one mass 'shot'. This amount is pre-calculated and must be accurate. The same pattern may not work with different moulds and should be worked out individually.

–   *Never back-track*: Make sure the mixed material (in liquid form) is spread gently at the bottom of a mould and not splashed, which will lead to product defects. In injection, it will lead to knit lines and possibly other problems.
–   *Follow the mould design*: The injection force/speed must ensure the smooth flow of material to cover all parts of a mould, which may have 'thick' and 'thin' areas. This means that the thick areas must receive more material and the thin areas less.
–   *Watch for tilted moulds:* If this happens, the injected/poured material will run 'downhill' first before filling the other areas. Remember, the time period between a pour/injection and creaming time is only a few seconds. The whole material mass must cream at the same time and not in sections.
–   *Pay attention to venting patterns*: Most moulds are 'feather-vented' meaning there is a slight clearance between the two mould halves running the whole way around the mould. When the material starts to foam, the air build-up inside the mould is pushed out through this small gap. If the mould has 'notched' vents, the air will escape through these. Depending on the design of these vents, there is a possibility that the foam can come out through these vents. With these types of moulds, injection should be done from the opposite end.

8.6.7 Using mould release agents

Mould release agents are an essential part in PUR mouldings. They have two functions in that in addition to easy removal of moulded parts, they also help in easy material flow and also good quality surfaces. Mould release agents in general do not affect the chemical mixture and most foam producers will be familiar with which ones to use for a particular application. The release agents can be applied by hand or better applied by spraying.

Here, an operator must exercise caution not to spray too much or too little as it will affect the moulded product. Generally, it may not be necessary to spray every time and once in three to four times may suffice. There are two ways in which an operator can detect that too much spray has been used: obvious small puddles after spraying and the parts are de-moulded. The other is small wet patches on the product surface or a hazy surface appearance.

8.6.8 In-mould coatings

In-mould coating uses special types of paint. Instead of painting a product after moulding, a coating is applied inside a mould before the PU mixture is dispensed into the mould. The moulded part comes out already painted. This sounds simple enough but it is not so. A foam moulder unless has enough experience should seek the assistance of the systems supplier or a chemical expert. Some of these special paints contain mould release agents and additional release agent coatings are not required. The principles of applying/spraying in-mould coatings are the same as for applying release agents.

8.6.9 Moulding with inserts

This is a very interesting and versatile aspect of PU moulding. Inserts can serve many different purposes. They can be support points, bases for screws and bolts. For furniture, plastic plates and metal rods often serve this purpose, while screw bases are good examples for moulding footwear, especially sports shoes.

The three basic things to remember when using inserts in a mould are as follows:

- *Temperature*: The inserts must be the same temperature as the mould temperature as cold inserts will result in sink holes and probable rejects.
- *Contamination*: The inserts must be completely clean from items like grease, oil, release agent or foreign matter which will weaken or destroy a strong bond between the inserts and PU foam.
- *Positioning*: The inserts must be held firmly in place as otherwise they will get 'bumped' due to the force of the rising foam.

8.6.10 Mould clamping

When PU foam systems react, they will create high pressure, more so, in a closed mould. The clamping force will naturally have to be more than the inside pressure. If the mould clamping force is insufficient, the two halves will open and the material will flow out. The same applies for uneven clamping force, with material leaking out from the weaker spots, thus causing 'flash' and material loss. Clamping forces can be applied either mechanically or hydraulically.

8.6.11 Mould cleaning

Over time moulds become 'dirty' from build-up of mould release and foam residue. Even before that the build-up on the surface of a part and also the tendency of making de-moulding was difficult, like the tendency to stick. Periodic cleaning will solve this problem but precautions must be taken if aggressive chemicals are used.

8.6.12 Flushing the mix head

It is an important phase of a production run. Use flushing agents or others to constantly clean the mix head to prevent any material build-up inside. Most machines will have a button to flush but advanced machines will have an auto-flushing function. Machines will have a drum or tank filled with a suitable flushing agent connected directly to the mix head.

8.6.13 Lead–lag when processing

Lead–lag will occur if the polyol and the isocyanate does not enter the mix head simultaneously. One component will lead the other which lags behind, creating poorly foamed areas. This condition will happen usually at the beginning of a production run. To do a lead–lag test carry out the following procedure:
-   Pour a thin layer of the mix onto a polythene sheet.
-   Allow the mix to foam and cure for a few minutes.
-   Peel the foam 'pancake' from the sheet.
-   Brush the bottom side of the foam mass with a 20% solution of hydrogen peroxide, which will indicate which component is lagging behind.

After brushing, one of the following can be observed:
-   Entire surface turns pale yellow indicating there is no lead–lag.
-   A dark-yellow stain appears around the initial shot point. This indicates isocyanate lead.
-   A white halo appearing around the initial shot point indicates a polyol lead.
-   The surface covered with alternating white and yellow streaks indicates poor mixing.

8.7 Troubleshooting

As in any manufacture, processing PUs will also encounter certain processing defects. Chemicals being involved with high exothermic reactions, the possibility of processing defects may be considered a little more complicated than processing standard thermoplastics. A foam manufacturer or moulder must have a good knowledge of possible processing defects and solutions in order to minimise waste and ensure good quality products.

8.7.1 Deficiencies in large foam block manufactures

Table 8.5 shows defects and solutions for processing large foam blocks.

Table 8.5: Deficiencies in large flexible foam blocks.

Defect	Description	Recommendations
Bottom cavitation	Bottom eaten away	Errors in metering – decrease tin catalyst
Dense bottom skin	Thick foam at bottom	Increase silicone level
Smoking	Excessive isocyanate vapours	Reduce isocyanate level
Tacky bun surface	Foam surface sticky too long	Increase total catalyst levels – errors in metering
Flashing/sparklers	Effervescence on rising foam	Decrease isocyanate – decrease silicone/amine – increase tin catalyst and reduce component temperatures
Friable skin	Skin flakes off at touch	Increase/change amine catalyst – increase component temperatures
Gross splits	Vertical/horizontal splits	Increase tin catalyst – decrease amine/water content
Heavy skin	Thick, high-density skin	Increase catalysts – increase isocyanate and reduce air entrapment in pour
Moon craters	Small pockmarks on buns	Reduce air entrapment in pour – minimise splashing on pour
Pee holes	Small spherical holes	Increase silicone – reduce mixing speed – reduce tank agitation and reduce air entrapment on pour
Relaxation	Foam rises and goes down	Check/increase tin catalyst – increase silicone level – reduce amine and reduce mixing speed/nucleation
Uneven foam colour	Colour concentrates in patches	Mix polyol well after introduction of pigment

8.7.2 Defects in moulding

Unlike solutions for large foam block defects, which may be considered fairly straightforward, solutions for moulding defects are a little more complicated. However, these will present exciting challenges to PU foam moulders and some may even have their own solutions which may be better than the ones recommended. Table 8.6 shows some of the most common defects and recommended solutions.

Table 8.6: PU moulding defects.

Defect	Description	Recommended solution
Air bubbles	Gas/air bubbles on surface	Improve venting – vents blocked
Short product	Incomplete product	Increase material – check injection port
Hard and soft foam areas	Varying densities	Insufficient mixing time
Knit lines	Lines in foam surface	Inject without back-tracking
White surface areas	Whitish patches on surface	Too much release agent – reduce and also clean mould
Material flash	Foam flash	Increase clamping force – make sure mould halves fit closely
Blisters on foam surface	Tiny bubbles on surface	Mould temperature too high
Difficult to de-mould	Moulded product sticks	Insufficient release agent – mould not clean

Bibliography

1. Nordson Sealant Equipment – *Meter-Mix Dispensing Systems for Polyurethane Elastomers* – www.sealantequipment.com/technical/polyurethane-systems.htm
2. Polytek Development Corp. – 'Technical Bulletin on Polyfoam Series' – www.polytek.com
3. BAUMERK Construction Chemicals – 'PUR IN 24' – Polyurethane based two component Injection resin – www.baumerk.com
4. COVESTRO Polyurethanes – 'How RIM Works' – www.polyurethanes.covestro.com
5. Hennecke Polyurethane Technology – 'TOPLine HK' – www.hennecke.com
6. Modern Enterprises (India) – 'Foaming Machinery' – www.foam-machinery.com
7. PREMILEC INC. – 'MAXFLEX 422' – Flexible moulded foam system – USA/Canada
8. Defonseka, C. – 'Practical Guide to Flexible Polyurethane Foams' – pages 128–137 – Smithers Rapra UK-2013
9. Edge-Sweets Company (ESCO) – 'ECD – Low-Pressure Dispensing Machines' – USA

9 Project set-up for small volume foam producers

The two-component polyurethane (PU) systems are generally not used for large volume foam productions, for example, in continuous foaming systems but can be successfully adapted for intermittent productions, meaning making single blocks at a time. The author will present an interesting project based on the standard principles of mixing and open pour for making large foam blocks where three different types of foams: (1) *standard flexible*, (2) *high-resilience* and (3) *viscoelastic foams* can be made on the same foaming machine system.

The following is a demonstration based on technical know-how accumulated over a number of years, on a hands-on basis and innovative designs particularly suitable for entrepreneurs or small volume foam producers with limited resources. Hope it will stimulate the minds of anyone interested in PU foam technology with the realisation that innovative designing can be applied to other branches of PU moulding with success.

What makes it really interesting is that this foaming and cutting system was designed and made from scratch by the author by using locally available parts (Canada) and using locally skilled personnel. This simple, yet, versatile system can easily be made by an entrepreneur or a foam producer already in manual operation and does not have to invest in expensive foaming and cutting systems. This concept was designed for an in-house operation and made for a local furniture manufacturer and ideal for use with two-component PU systems available in drums or in totes (larger volumes).

9.1 The complete system

This project was set up in a 4,000 square foot building with three-phase power and 15 A wiring. The basics of the operating system was that the two components in drums were directly connected to the large steel mixing cylinder with mix head, which moved up and down vertically on both manual and auto mode. The two moulds made from laminated board with a base on wheels and all four sides detachable were used for production. An exhaust system was also installed. The whole system was controlled by an electronic control board which included auto metering of components, mixing ratio variations, auto-mixing and cut-off, auto-lifting of mixing cylinder and other features. The system was protected by an emergency stop button. Figure 9.1 shows the full system.

https://doi.org/10.1515/9783110643169-009

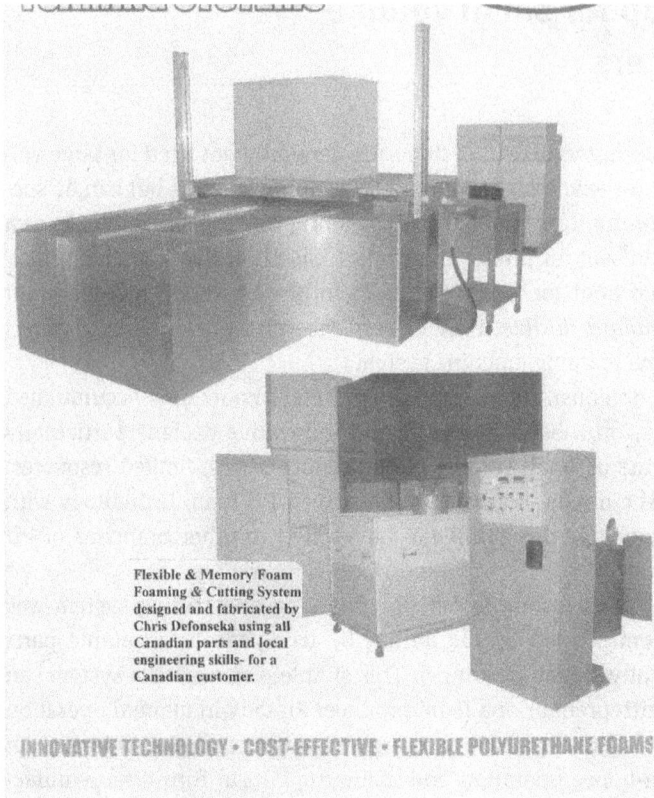

Figure 9.1: Full foaming and cutting systems.
Source: Courtesy of the author.

9.2 Raw material systems

A leading PU systems supplier in Montreal was contacted through their local head office in Toronto. They were already into providing two-component PUR systems for coating and rigid applications but were willing to supply two-component systems for flexible and memory foam productions. They had well-equipped laboratories but did not possess technology for formulating flexible and viscoelastic foams (memory foam). The author provided suitable formulations and under his guidance, suitable two-component systems were formulated with three different densities for flexible foam productions and one density for memory foam. For easy identification, colour codes for all materials provided was also agreed upon. The price per kilogram of materials included delivery to the factory premises in Toronto, based on minimum quantity purchases. The shelf-life was six months at maximum storage temperature around 25 °C (77 °F). For high-resilience foam production, a special grade – Greenlink HR 250 two-component system from ERA Polymers Ltd. to be used.

9.3 The foaming process in brief

The mould on wheels is brought directly under the mixing cylinder. Since the pour mix will be in liquid form and very difficult to stop leakage at the bottom, an innovative concept was used in that a very thin sheet of flexible foam of size – 45 cm × 45 cm × 1.25 cm (thick) is embedded in the centre of the mould floor covered in a polythene sheet, flushed with the surface. The mixing cylinder is lowered onto this foam base and its weight makes it 'sink' into the foam surface, thus preventing any leaks. Another alternative is using a suitable gasket but most will deteriorate soon and will have to be replaced often.

The desired amounts of components A and B and the mixing ratios and times are now set on the control panel. Component B (polyol blend) is now metred into the mixing cylinder and if any colour is added, mixed for a few seconds. Component A is then metred into the mixing cylinder containing the polyol and the mixing mode is activated. In a few seconds, as per time set, the mixing cylinder will automatically lift vertically, releasing the foam mix still in liquid form, to spread evenly to cover the whole mould floor. The liquid mixture will turn 'creamy' and start to rise slowly.

The operators must wear safety equipment and avoid the natural tendency to look into the mould as the rising foam will be giving off toxic gases. When the foam has risen about one-third of the height of the mould, a very light 'floating lid' will be placed on top of the foam to prevent a meniscus (rounded top) being formed. Until the foam is semi-cured (surface not tacky) and firm, the mould should not be moved. De-moulding can take place after about 10 minutes and the

foam block should be kept upside down when it is still slightly warm to flatten the surface further to minimise the waste. Keep in a holding area until ready to be moved to the final curing area using the first-in–first-out (FIFO) system. Figure 9.2 shows moulded foam blocks kept upside down in a holding area. Note the 'whitish patch' on the foam block, which is the pour point at the bottom.

9.3.1 Foam cutting system

This is a long sturdy frame made of angle iron and horizontally moving flat bed, motorised with variable speed-gear motion. Movement includes both forward and reverse. Two vertical arms with long open grooves to enable the hotwire/wires to be moved up and down or angled cutting with single wire or multi-wires' as desired. A simple movable control box on wheels housing a 30 A inverter to provide a voltage range: 10 V to 100 V – for heating the spring-loaded Ni Cr cutting wires fixed across the two vertical aluminium arms. The voltage can be adjusted according to the number of wires used. This cutting machine can trim all sides of a foam block and if multi-wires are used, it can cut foam slabs of different thicknesses at th same time. If hotwire cutting is not allowed according to local municipal laws, an

Figure 9.2: Moulded foam blocks in holding area.
Source: Courtesy of the author.

alternative would be a band saw machine, in which case both horizontal and vertical cutting machines will be needed.

The following manufacturing procedures may be used as examples of processing methods of three different qualities of PU foams using two-component systems:

9.4 Manufacture of flexible foam blocks for mattresses

Most of us are already familiar with the processing methods for flexible foams for mattresses using formulations with multiple chemicals and hence, the author will only emphasise on the processing aspects of using two-component systems. The use of simplified two-component systems, where the isocyanate is supplied as component A and a blended polyol as component B, makes processing that much easier. However, the disadvantage is that the foams produced will only have the formulated density and properties per grade. By varying the proportions of A and B, the density may be increased or decreased a little but within reasonable limits.

9.4.1 Making a flexible foam block for 'Queen-Size' mattresses

In earlier days, a foam mattress meant a slab of foam 4 inches (10 cm) thick with a cloth cover but not so now. As the demand for comfort levels have increased and a variety of materials are available, a mattress means, a mattress built up of several layers. Today basic mattress sizes are as follows:

- Standard: 36 in. × 72 in. × 4 in.(90 cm × 180 × 10 cm)
- Double: 39 in. × 75 in. × 4 in.(97.5 cm × 187.5 cm × 10 cm)
- Queen: 60 in. × 80 in. × 5 in.(150 cm × 200 cm × 12.5 cm)
- King: 76 in. × 80 in. × 6–8 in.(190 cm × 200 cm × 15–20 cm)

Note: Actual sizes may vary with different manufacturers

9.4.2 Production specifications

This presentation is about manufacturing queen-size mattresses of dimensions as shown above. A large flexible foam block will be made from which the required foam slabs can be cut. For this exercise the processing specifications are as follows:
- Mattress size: 150 cm × 200 cm x 12.5 cm
- Number of mattresses per foam block:8
- Density of foam: 27.2 kg/cu.m
- PU System: Two-component
- Processing parameters as per Table 9.1
- IFD: 2.3

Table 9.1: Processing parameters.

Component A	40 pbw
Component B	100
Mixing ratio	100:40
Mixing time (seconds)	8
Cream time (seconds)	12
Gel time (seconds)	70
Tack-free time (seconds)	300

9.4.3 Calculating size of foam block required

Final product: 8 mattresses of size 150 cm × 200 cm × 12.5 cm per block

Calculation in inches = 60 in. × 80 in. × 40 in. = 61 in. × 81 in. × 41 in(with allowances for trims)

Now apply, $M = V \times D$ where M= amount of material V = volume D = density
Then, $M = 61 \times 81 \times 41$ (117.23 cu.ft.) × 27.2
   = 7.33 cu.m × 27.2 kg per cu.m
   = 199.38 kg + 1 % gas loss + 3% material waste = 207.36 kg

Therefore, each pour should be = 207.36 kg made up of
  Component A = 40 ÷140 × 207.36 = 59.25 kg
  Component B = 100 ÷140 × 207.36 = 148.11 kg

9.4.4 Processing method

Apply one or two coats of a release agent on the inside smooth surfaces of the assembled mould. If application is by hand, it may be easier to apply the coats on the panels before assembly. Bring the mould on wheels directly under the mixing cylinder and centre it. Lower the mixing cylinder to sit tight on the floor and secure it to prevent leaks. Before connecting the raw material drums to the machine, stir the contents for a few minutes at slow speed.

Now, set the metering and mixing parameters on the control panel. The metre X component B (polyol) into the mixing cylinder. If any colour or other extra additive is to be added, do so now and mix the polyol blend for 20–30 seconds on manual mode. Component A can now be metred into the mixing cylinder and activate auto mixing which will raise the mixing cylinder automatically, releasing the foam mixture to spread on evenly to cover the whole mould floor. After a few seconds, the mixture will turn creamy and begins to rise slowly. When the foam mass has risen about one-third the height of the mould, place a very light floating lid on the rising surface to stop a rounded top being formed. The mould should not be moved until the foam block is not sticky and firm. After de-moulding, keep the foam blocks upside down in the holding area until removal to the final curing area. Foam blocks will be undergoing exothermic (heat giving) chemical reactions and in order to prevent fire hazards, keep the blocks one foot apart and do not stack them one on top of another. Full curing will need at least 24 hours.

9.4.5 Fabrication and cutting process

Hotwire cutting with single wire or multiple wire cutting will give a very smooth surface cut with negligible material loss when thin gauged wires are used. For large volume foam block cutting, band saw machines are preferred and a small allowance for material loss must be allowed due to the thickness of the cutting blade. It is advisable to have at least one vertical and one horizontal cutters with the first operation being the trimming of all sides, before cutting the desired thickness for the mattresses. The foam wastes can be recycled and large volumes can be re-bonded using steam/compression or adhesive/compression methods and made into large blocks and then cut into thin sheets for carpet underlay, foam slabs for mattress bases or other.

Figure 9.3: Multiple Foam Block Cutting Machine.
Source: Reproduced with permission from Modern Enterprises, India.

Figure 9.3 shows a multiple-block band saw cutting system. All blocks can be cut into desired thicknesses for mattresses on auto-mode, applicable for large volume productions.

9.5 Manufacture of high-resilience foam blocks for office and domestic cushions

High-resilience PU foams are made from special grades of polyol mixtures and provides ultra-comfort and may be considered as a higher grade of standard flexible PU foams. Whether these materials are used in mattress applications, cushions, mattress toppers or other applications they provide extra comfort and good support. In density, they will range higher than 40 kg/cu.m and their cell structures enable extreme elasticity and optimal supporting forces. These foams distribute pressure across an entire surface and will block the transfer of motion from one sleeper to another in mattress applications, while providing extended extra comfort as cushion seat, especially for prolonged seating.

The basic difference between regular flexible PU foams and high-resilience (HR) foams is found in the chemicals used to make these foams. HR foams use higher quality chemicals that contribute to elasticity and longer comfort life. HR foams are heavier than normal flexible foams due to higher densities and will cost more but users do not mind the extra cost due to its added superior comfort and durability. Perhaps, a compromise would be a combination of a bottom layer of flexible foam and an upper layer of HR foam. In a way, HR foams may be compared closer to memory foams in that their surfaces are soft and yielding but unlike memory foams, indentation recovery is much faster. There are many suppliers of two-component HR systems with different grades, each with speciality functions but in principle, the basic functions are the same.

9.5.1 Speciality HR grade

Greenlink HR 250 from ERA Polymers Ltd. is a high-resilience water-blown flexible two-component PU system. This product is specially designed for the manufacture of cushions for office and domestic furniture. Although, this system can be manually drill-mixed (for small blocks) at a minimum speed of 3,000 rpm, for large blocks a machine must be used to handle the large volume of mixture.

Specifications for processing as shown by Table 9.2

Table 9.2: Specifications as per data sheet – ERA Polymers Ltd.

Description	Polyol	Isocyanate
Appearance	Opaque liquid	Brown liquid
Viscosity (cps)	1450	450
Specific gravity	1.05	1.20
Isocyanate value NCO (%)	N/A	25.5
Mix time (seconds)	7–8	–
Cream time (seconds)	12	–
Gel time (seconds)	70	–
Tack-free time (seconds)	300	–
Free-rise density (kg/cu.m)	57	–
Mixing ratio	100:57	–

9.5.2 Production data

Product: Large HR foam block from which different size cushions can be cut.
Cushion size (standard): 22 in. × 22 in. × 5 in(thick) – (55 cm × 55 cm × 12.5 cm)
Foam block size needed to cut 32 cushions = 44 in. × 44 in. × 40 in.(110 cm × 110 cm × 100 cm)
Actual foam block to be produced before trimming: 45 in. × 45 in. × 41 in. – (112.5 cm × 112.5 cm × 102.5 cm) with allowances for trims.
Density: 57 kg/cu.m

9.5.3 Mould requirements

Moulds can be fabricated from a variety of materials such as metal, aluminium, fibreglass and thick plastic sheets but recommended material would be 2.5 cm thick laminated wood (one side smooth) keeping in mind the costs as well as the forces exerted on the mould sides due to the tremendous pressure build-up during foaming. All sides should be detachable and for easy movement, the mould base should be on wheels. To condition a new mould before use, two coats of a release agent, such as Eralease can be applied and the surfaces 'polished'. Unlike flexible foams, adhesion to mould sides will be less when processing high-resilience (HR) foams due to its silky surfaces. The use of a 'floating lid' made of a very light material will help greatly in reducing foam waste on the top surface.

9.5.4 Calculating material required for pour per block

To calculate the total amount of material required per pour, follow these steps:

Apply formula, $M = V \times D$ where, M = material amount in kg V = volume D = density

Then, M = 45 × 45 × 41 cu.in. × 57 kg/cu.m
$\qquad M$ = 1.36 cu.m × 57 kg/cu.m

Therefore, M = 77.52 kg + 1% gas loss + 3% material waste = 80.62 kg

The total material per pour = 80.62 kg
Now, component mixing ratio recommended by system supplier = 100 B: 57 A

Then, Component A = 57 ÷157 × 80.62 = 29.27 kg
$\qquad$ Component B = 100 ÷157 × 80.62 = 51.35 kg

9.5.5 Processing method

On the control panel board, set component A to 29.27 kg and component B to 51.35 kg. Set mixing time to 8 seconds with auto-lift. Since the controls are calibrated and works on timers, the settings must be accurate. Bring the mould on wheels under the mixing cylinder and position it at the centre and lower the mixing cylinder onto the floor of the mould and secure it. Meter component B into the mixing cylinder. If any colour or additive is to be added, do so now and mix for 20–30 seconds on manual mode. Set mixing on auto mode. Meter component A by activating switch and immediately activate mixing on auto mode. After 8 seconds, the cylinder will rise vertically slowly releasing the foam mixture still in liquid form to spread and cover the whole mould floor. The mixture will turn a 'creamy' colour and begin to rise slowly. When it has risen about one-third of mould height, place the floating lid on the surface of the rising foam. This lid must be very light but also have sufficient weight to suppress the curved top (meniscus) formed on the rising foam surface. When the foam has risen fully and settled down, remove the floating lid but do not move the mould until the surface is not tacky.

In about 10 minutes, the mould can be wheeled out and de-moulded. The still warm foam block can now be kept upside down in a holding area until removed to the final curing area, where the foam blocks should be kept at least one foot apart.

Do not pile one foam block on top of another. Exothermic (heat giving) reactions are taking place inside the blocks and good ventilation and an exhaust system is needed in this area. Foam blocks can be taken for fabrication/cutting after a recommended minimum curing period of 36 hours.

9.5.6 Cutting of foam blocks

The current common practice is to use manual, semi-auto or automatic band saw cutting machines. In auto-cutting, several blocks can be loaded and cut into desired thicknesses at the same time. With contour cutting machines, different shapes can be cut.

9.6 Manufacture of viscoelastic/memory foam blocks for mattresses

A viscoelastic foam is a flexible PU foam specially formulated to increase its density and viscosity and unlike standard PU foams it has low resilience and pressure-sensitive properties. Unlike standard PU foams it has four dimensions: (1) *hardness*, (2) *density*, (3) *temperature and* (4) *time* – reacting to body pressure within these zones and with a 'lazy' or slow recovery with high damping. These foams have very low resilience, allowing a body to sink into the foam taking the full contour of a body without pressure, irrespective of shape or size and cushioning the full body giving a sensation of floating on a cloud.

Viscoelastic foams are of greater densities than standard flexible PU foams. Densities of ≥80 kg/cu.m are considered as 'high-end' quality, whereas densities in the range of 64–72 kg/cu.m are the more common ones. Most viscoelastic foam mattresses are from IFD 12 to 16. In terms of densities rated in lbs.cu.ft. mattresses with densities 4–5 lbs/cu.ft. are acceptable.

Some manufacturers of mattresses use 8 inches (20 cm) thick slabs but generally, a 4 inches (10 cm) thickness should be sufficient. Since viscoelastic foams are very expensive, some may opt to use a combination such as 2 inches (5 cm) recycled slab base + 3 inches (7.5 cm) standard foam + 3 inches (7.5 cm) viscoelastic foam or even use a 2 inches (5 cm) or 3 inches (7.5 cm) mattress topper of viscoelastic foam on top of an existing mattress. Since these mattresses are very heavy, manufacturers will use vacuum packing for marketing purposes.

9.6.1 Mattress size

This exercise will deal with the manufacture of viscoelastic foam queen-size mattresses. On the basis of a single mattress being – 80 in. (200 cm) × 60 in. (150 cm) the required moulded foam block would be 81 in. (202.5 cm) 61 in. (152.5 cm) with a height of 25 in. (62.5 cm) with allowances for trims. From the trimmed block, 4 in. (10 cm), 6 in. (15 cm) thick or other can be cut to complete as mattresses. Unlike flexible foams, where the moulded height of a block can be 42 in. (105 cm), which is standard even for continuous slabstock foaming, the maximum height recommended is 28 in. (70 cm) with 24 in. (60 cm) being a convenient and practical height due to heavy weight of foam. Foam blocks with heights over 28 in. will be very heavy and difficult to handle.

9.6.2 Mould construction

Mould construction can be more or less as already discussed but the bottom panel (base) should be about 2.5 cm thicker to bear the extra-heavy load and on wheels for easy movement. The insides of the mould should be smooth and assembled inside volume of the mould should be 202.5 cm × 152.5 cm × 75 cm. Depending on the grade of two-component system being used, shrinkage may be less than for flexible foams. Here, a pre-production lab box test will be very helpful. The use of a standard floating-lid will help to reduce foam waste at the top surface.

9.6.3 Calculating material per pour

Material: two-component viscoelastic
Mixing ratio: component A: component B = 1:1
Density = 5.0 lbs/cu.ft.
Volume per block = 81 in. × 61 in. × 25 in. = 71.48 cu.ft.
Applying $M = V \times D$ where M = mass V = volume D = density

$$M = 71.48 \text{ cu.ft.} \times 5.0 \text{ lbs/cu.ft.} = 357.4 \text{ lbs} = 162.45 \text{ kg} + 1\% \text{ gas loss} + 3\% \text{ waste}$$
$$= 168.96 \text{ kg}$$

Therefore, component A = 84.48 kg component B = 84.48

9.6.4 Processing method

It is very important that component B is mixed well before being used and also the final mixing time otherwise the quality of the foam will be affected and can be detected only after cutting. When this component is metered into the mixing cylinder, the desired colour may be incorporated. Fillers may or may not be used but if yes, instead of standard calcium carbonate, use 10–15% of rice hull powder or bamboo powder of particle size ≤5 microns which will have better properties.

Same is for flexible and HR foams. It may be difficult to turn these post-moulded foam blocks upside down as for others but recommended solution is to apply slight even pressure by placing light weights or other on the surface of the floating lid to flatten the warm surface in order to reduce the material waste. As a rule, viscoelastic foams need longer curing times – 48 hours at least before cutting into desired mattress sizes.

9.6.5 Important quality checks

Since these are expensive high-end products, it is recommended that the following quality checks are carried out:
-   IFD: It should be much higher than for flexible or HR foams as per market demand. This can be measured on the production floor with a small hand-held meter.
-   A mattress when bent in two should not have any splits.
-   A viscoelastic mattress topper – 2 or 3 in. thick, when rolled up should not have any splits.
-   Check indentation recovery time as per market requirements.
-   Cut a small piece of foam – 10 in. (25 cm) × 2 in. (5 cm) and freeze it for one hour. For example, use 'freezer-box' of a refrigerator. If the foam becomes frozen-stiff, the viscoelastic quality is very good. You will see that a similar piece of flexible foam will not be affected by freezing.
-   Cut a full-size mattress 6–8 inches thick (15–20 cm) and keep a glass full of liquid on one side of the surface and if somebody walks on the other side, the glass will remain steady without spilling any of its contents. This is a good marketing tool in addition to the standard requirements of density and IFD.

Because of the heavy weight of a mattress, they are vacuum-packed (size reduction) in a thick plastic bag and put in a cardboard box for easy handling and transport. It is recommended that a 'perfume-bouquet' or something similar is put inside the packing as when the pack is opened the viscoelastic material will give off a slight odour which can last a few hours.

Bibliography

1.   Defonseka, C – 'Practical Guidelines to Flexible Polyurethane Foams' – Smithers Rapra UK– 2013 pages 130–135 pages 8–11 pages 83–90.

10 A case study: reduction of excessive waste, improvement of foam quality and process efficiency

Introduction

The manufacture of polyurethane (PU) foam though a lucrative business and serves vital needs of people, especially comfort, footwear, medical, automobile and so many other areas of applications also have to overcome disadvantages such as wastes, safety factors, quality issues and so on, especially when very large production volumes are involved. Generally accepted foam waste factors can be as high as 15–20%, with some of the foam recyclable.

Take the case of a large manufacturer of flexible PU foams dealing with products for comfort applications such as mattresses, cushions and sheets. This organisation consisted of a modern factory premises around 200,000 square feet (square metres) employing 320 personnel. All products were made to ISO standards and marketed under a well-known brand. Their manufacturing team consisted of a technical director, production manager and heads of various departments. The company had started in a small way and over the years had steadily expanded.

The plant consisted of a continuous foaming system supported by manual, semi-auto and automatic cutting systems. For recycling of foam wastes, an efficient re-bonding machine was available. The morale of the working staff was good but with expansion and increasing production volumes the company began to have predictable manufacturing problems with regard to quality issues, excessive wastes and declining process efficiency because of lack of technical know-how, naturally resulting in drop-in profits. The board of directors had sought assistance of two local consultants but since they were unable to produce any satisfactory results, decided to seek overseas help and the author's services were sought as a consultant.

10.1 Analysis of Concerns

This company had started as a small single owner operation to make PUR foam mattresses which expanded rapidly due to demand. Diligent planning at the beginning had made available plenty of land space for expansion and during their 20-year operation as their brand name established itself with increasing sales volumes, this company was able to slowly expand with periodic factory floor additions.

https://doi.org/10.1515/9783110643169-010

Since their production methods were more or less manual, they had decided on a fully automatic continuous foaming line and suitable cutting and fabricating systems. Foaming consisted of two short runs done in the afternoon. Their concerns started at this stage because of the following reasons:
- Loss of foam waste because of curved surface on foam 'buns'
- Different qualities of foam from same formulation
- Foam produced for sheeting for peeling machine was too coarse (sticks on cutting blade)
- Excessive adhesion of hot foam on sides of 'paper trough' on foam conveyor
- Sometimes foam blocks collapsing when new formulations are used
- Cut foam buns from foaming line sent to final curing area and stored at random pattern
- Lack of quality checks for foamed buns before storage

10.2 Overall concerns

On the basis of overall concerns covering production, processing, assembly and marketing of products, the following were highlighted by the customer for solutions:
- Total foam waste: Current 33% as against standard acceptable 15%
- Process efficiency: Low at 40% and increase to 75%
- Marketing problem: IFD (support factor) less than 2.0 (minimum) – increase
- Cutting: Large slabs left over and sent for recycling – reduce excessive waste
- Fabrication: Introduce QA system to reduce/eliminate rejects

10.3 Pre-planning of assignment

In order to work out an efficient work plan, the author (as consultant) commenced a dialogue via phone calls and emails and worked out the following tentative work schedule to be fine-tuned on his arrival at the factory:

Team:	Consultant
	General manager (working contact)
	Project secretary
	Project coordinator
Work schedule:	Assignment duration: 4 weeks
	Assignment period: March/April 2010
Assignment targets:	1. Reduce foam waste from 33% to 15%
	2. Improving foam quality issues
	3. Increase process efficiency from 40% to 75%
	4. Increase foam IFD form less than 2.0 to minimum 2.2

Procedure methodology:

Week 1 – Introduction to board of directors and management – inspection of factory – brief study of current operations – examination and evaluation of company financial, production, marketing and other relevant records and data – discussions with directors – meetings with key departmental heads and personnel – selection of final work team consisting of consultant, general manager, personal secretary, production manager, engineer, technologist and coordinator. The consultant would work with each department and conduct daily meetings with full team for discussions and implementations of recommendations.

Week 2 – Ongoing study and research with close observation and recording of each phase of activity and overall production flow – from foaming to storage of foam blocks to cutting and fabrication to assembly and shipping.

Action flow

- 8 a.m. – 1 p.m. (observation, floor discussions and recordings)
- 2 pm – 5 pm (training of personnel of key departments)

10.4 Training parameters and methodology

Parameters: Basic polymer and PUR technology – functions of components – foaming calculations and formulating techniques – troubleshooting foam defects – monitoring – recording – evaluation and action – quality control systems for PUR foams – statistical process control (SPC) – cutting techniques – minimal waste/zero waste cutting – how to set up an in-house laboratory – ISO/ASTM standards – preventive maintenance methods – introduction to log books at strategic stations – safety factors – chemical spill management.

Methodology: Individual and group discussions of problem areas – group lectures – videos – handouts – literature.

10.5 Overall production process as observed

All chemicals are purchased in bulk and on arrival are pumped into large holding tanks which are connected to the main foaming machine. A crew of five operators with one lead-hand was in charge of this foaming line. Being a modern automatic foaming system capable of came in at 7 a.m. but arrived at the machine only at 8 a.m. and the setting up took 2 hours and was ready for a production run only around

10 a.m. The lead-hand then studied the production schedule for that day as issued to them by the production department and was seen discussing it with his team.

The requirement for that day was two runs of 700 kg each but of two different densities. The chemical metering controls were then set by the lead-hand and the mixing head deposited a small quantity of the foam mixture on to the floor of the slowly moving conveyor carrying a paper trough. This under-mixed foam mixture was slow to rise and formed an uneven mass about 20 inches in height. This was an inherent initial waste and could be considered normal.

The subsequent flow was good and rose to the pre-determined height of 42 inches (105 cm) and forming a 1.5 inches (3.75 cm) meniscus (curved top). As the foam mass moved forward it began to 'cure' forming a solid mass but very hot because of the exothermic reactions in the foam. Down the line the foam mass cooled down somewhat and a vertical cutting system cut the foam into large standard size 'buns'. An operator marked the density and date on each block and then kept them in a holding area until transporting to the final curing storage. It was observed that there were foam wastes at the beginning and at the end (normal) which when cooled, was sent for recycling. The entire foaming run took only 48 seconds.

The machine was then set up for the second run of a different density, with production parameters more or less the same. The time of commencement of the second run was at 11.30a.m. and the temperature was much warmer. It was observed that at the end of each run, the cut foam blocks were marked showing the date and density of each foam block. The crew came in then set about tidying up the foaming section. At 12.30 p.m. they went for their lunch break and returned at 1.30 p.m. and there was plenty of idle time, until at 3 p.m. when their shift ended. An important observation made was that while all the foam blocks made around 10 a.m. were of 'good foam quality', some of the foam blocks made during the second run around 11.30 p.m. was of poorer quality (foam block height variation and cell structure not uniform). (See Figure 10.1.)

Figure 10.1: PUR continuous F/ Line.
Source: Reproduced with permission from A.S. Enterprises Ltd., India.

Two production runs of 700 kg per day was the norm with the inherent wastes A at the start and D at the end. By weighing these and each block and using the chart above over a period of a week, it was established that 10% of the foam waste could be attributed to this. Another 3% could be due to occasional foam collapse, excessive foam adhering to the paper trough sides.

10.5.1 Recommended new procedure

The marketing section to work closely with the production section and work out product
requirements well in advance enabling them to work out a weekly production schedule and provide it to the cutting and fabrication department in advance. They will then work out a daily foaming schedule based on density/block size needed and the exact block lengths to produce (height remains the same). The current standard practice of cutting foam blocks to same lengths on the machine to be discontinued and foam blocks to be cut to different lengths as per dimensions worked out by the cutting section. This will minimise foam wastes during cutting and fabrication.

The foaming crew to come in at 7 a.m. is a normal practice but must carry out the first production run by 8 a.m. and the second foaming run by 10 a.m. when the atmospheric temperature is cooler. This will ensure good quality foam. To do this, soon after the second run is completed, the crew must clean the machine and set it up again for production before their shift ends. To overcome the problem of excessive adhesion of hot foam to the paper forming the 'foaming trough', replace the current paper being used with 'peel able – Kraft paper' which is a special grade of paper unglazed or unbleached but with a low or high density film which sticks to the foam without any adhesion of foam to the paper, resulting in smooth surfaces and less trimming.

Instead of a 48 second foaming run for 700 kg, the foaming run to be extended to 60 seconds, where there will be an increase of 175 kg (total 875 kg) while the foam wastes at the start and end (Figure 10.2. – A and D) remained the same. This means foam wastes as a percentage will come down significantly. This increased production will mean lesser number of 'foaming days'.

The cut foam blocks, while still warm to be kept upside down in the post-foaming holding area until removal to the final curing storage to suppress the curved top surface, thus reducing the 1.5 inch (3.75 cm) meniscus to 1.0 inch (2.5 cm). After discussion with the engineering section, a design was given to them to implement a light free-rolling horizontal cylinder which is to be applied on the top surface of the foaming stream to eliminate the curving of the foam on top. This will save a considerable amount of foam (3.5%) and when this is implemented the formulations could be adjusted accordingly.

Weekly Waste Check-Foaming Line

To: Chief Executive Officer
      Operations Manager

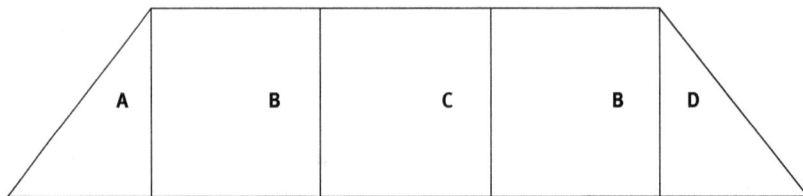

Day	Total Raw Material Used	No. of B Blocks	No. of C Blocks	Waste A	Waste D	Total Weight of Blocks (kgs)	Total Waste kgs.	Waste %
2								
4								

_____   Prepared By:_____

Figure 10.2: Machine foaming pattern and chart to calculate weekly foam wastes.
Source: Figure 10.2: A chart designed by the author
A = foam waste at start of production run
B = foam blocks cut to required sizes to minimise waste in fabrication
C = foam blocks cut to required sizes to minimise waste in fabrication
D = foam waste at the end of production run

All foam blocks after cutting to be weighed on the platform weighing machine available at the foaming machine. Each block to have a tag with data such as weight, density, date and dimensions (mostly the length) and formula code. A daily

detailed report of production to be sent to the production manager. Under the com-
ments column, a short report of the day's work done would greatly help.

10.6 Post-production storage

The post-production storage area for foam blocks was very good and clean, with
efficient ventilation and gas-exhaust systems in place. The foam blocks were placed
in a 'mixed' random pattern during the post-curing stage of 24 hours. During this
phase, gas is emitted from the exothermic (heat giving) reactions still taking place
inside the foam blocks. Hence, the need for good ventilation and exhaust systems.
If by mistake, under-cured foam blocks are taken for cutting and fabrication, the
foam quality will be poor and also the possibility of fire hazards must be kept in
mind. This was observed in some blocks taken for cutting. When the top was
trimmed to remove the curved surface, the foam was still 'wet and sticky' and the
foam had to be rejected.

10.6.1 Recommendations

The storage of foam blocks should be on the first-in–first-out (FIFO) system and
stored at least one foot apart. Blocks should never be stacked on top of each
other.

10.7 Cutting and fabrication

The current practice, as observed was the trimming of the 1.5 inches (3.75 cm) thick
rounded top of all foam blocks taken for fabrication. Then all four sides and the
bottom were trimmed. Although, this was the normal current practice, since the
foam blocks on the machine were cut to one standard size, when mattresses and
cushions were cut, large slabs of good foam were left over and sent to the recycling
section. This was unacceptable.

   After weighing all foam leftovers/wastes per shift which included all trimmings
and post-fabrication over a 10 shift period and taking an average, it was concluded
that the cutting and fabrication section generated the most waste (around 20%) of
the overall 33% waste projection per month. This being the pivotal processing or
waste-producing action for the company's overall operation, it was surprising that
quality control procedures were non-existent. This was the section that should have
given the production manager a daily or at least a weekly report but none was available. Since no attempts were made to check the densities or the IFD of the foams,
these emerged only at the time of marketing, resulting in complaints by customers

and restraints in selling products. One additional problem observed was the occasional dimensional irregularities of foams cut for final assembly. These involved both mattresses and cushions, while the sheets were acceptable.

10.8 Quality control system set up

As we know, quality control is an essential function of any manufacturing activity. Generally, for individual customers, a company or a market will provide acceptable quality standards with tolerances as per local or international standards and manufacturers will have to meet them for successful marketing. In this case a simple quality control system was set up with training for two outside individuals as QC assistants for starters. For any QC system to be successful, the checks for quality should be carried out by independent individuals and not by section personnel. The QC system used was statistical process control (SPC), where parameters were set up for lower control limit (LCL) and upper control limit(UCL) and median. Control charts like X–R chart (for recording variable data) and P-chart (for recording attribute data) were also introduced, which would yield valuable information for improving and controlling quality. These information would be very valuable for the foaming section, the start of the whole production operation.

These charts would be used for the following important areas for producing quality products of the three main items: (1) mattresses, (2) cushions and (3) sheets.
– Density
– Support factor (IFD)
– Dimensions

In PU foams, density variations can occur and should be minimised to ensure maximum comfort and also prevent use of excess material, among other factors. Densities can be easily calculated using volume and weight data. IFD can be measured on the floor with a hand-held device, while dimension irregularities can be prevented with extra care during cutting of the foam. The introduction of suitable wooden frames for random checking would eliminate this problem.

A colour code system was introduced withgreen (acceptable), yellow (on hold) and red (reject). All products acceptable would carry a 'green sticker' and only these would be sent to the finishing section. Speciality products like wedges, medical seats and sponges would be cut from the products in the yellow bin and the leftovers to go into the red bin and sent to the recycling section. A daily report to the production manager was introduced for analysis and action where necessary.

10.9 Assembly and finishing

This section worked efficiently with the main products being foam mattresses. The work basically comprised assembly of foam slabs, padding, laying of foam sheets, soft patterned cloth and trimmings. However, they had two problem in that (1) dimensional variations in foam slabs and (2) sticking of the thin foam sheets on the cutting blade when cut on the peeling machine giving uneven and damaged surfaces. It was found that this was because of the foam sheet being too coarse when cut from the foam blocks made in the factory. They had to purchase suitable foam sheeting in roll form from and outside sources. Figure 10.3 shows a peeling machine where thin foam sheets are cut for padding.

Figure 10.3: Foam sheet peeling machine.
Source: Reproduced with permission from Modern Enterprises, India.

10.9.1 Recommended solutions

– *Dimensional variations: solution lay in cutting and fabrication section. See quality control system set up.*
– *Foam sheeting (peeling machine) too coarse: solution lay in adjusting the formulation. The foam structure was successfully converted from a 'coarse' to fine-cell structure by increasing air content by 1.0% and increase in blowing agent by 0.8%, all other parameters remaining the same.*

10.10 Process efficiency

The overall process efficiency of the plant was low mainly due to disjointed production flow and too much non-productive time. Each section worked within themselves and not creating a smooth flow of work. Machinery maintenance was poor, foam block storage system was a random pattern resulting in semi-cured blocks being taken for cutting and fabrication resulting in waste. Lack of quality control at this section was a big problem. Assessment of low process efficiency was lack of technology, unawareness of solutions to problems and inability to identify and implement effective solutions for efficient process flow. Corrective actions were implemented and the results would be seen only in due course. To enhance the success for improving production efficiency, it was recommended that a committee headed by the general manager/production manager and section heads meet regularly for efficient coordination of work.

10.11 Recycling section

Recycling system used was a size reduction/adhesive/high-pressure machine making large foam blocks which were cut into sheets (carpet underlay) and thick slabs for mattress bases. This section worked efficiently.

10.12 Marketing

The company products had established a strong market demand with a well-known brand name. However, some returns were experienced due to the support factor (IFD) being less than the 2.0
being below the required value and also due to dimensional irregularities.

10.12.1 Solutions implemented

- *Increase of IFD from 1.8 (current) to ≥ 2.2: increase in filler content by 10% and reduction of blowing agent by 0.5% in the formulation.*
- *Dimensional irregularities: for solutions see under cutting and fabrication section quality control procedures set up.*

10.13 Conclusion

The last two days of the assignment was spent in preparing a report for the board of directors and appraising them of the work done and the work which needed to be implemented. A final meeting with the heads of departments and management concluded the assignment.

Appendix 1 Some suppliers of two-component polyurethane systems, dispensing and cutting machines

A 1.1 PUR systems

- ERA Polymers Pty Ltd. – Australia
- EPIC Resins – Palmyra WI
- Normac Adhesive Products Inc.
- Baumerk Construction Chemicals – Turkey
- Northstar Polymers LLC – USA
- BASF Polyurethanes Asia Pacific – Hong Kong/India/Taiwan
- Polytek Development Corp. – USA
- Premilec Inc. – USA
- Bio Based Technologies LLC – USA
- Huntsman Corporation – USA
- Freeman Manufacturing & Supply Co. – USA

A 1.2 Dispensing machines

- Hennecke Gmbh – Germany
- Canon Viking – UK
- Edge Sweets Company (ESCO) – USA
- AS Enterprises – India
- Modern Enterprises – India
- Nordson Sealant Equipment – USA
- ERA Polymers Pty Ltd. – Australia (Agents)

A 1.3 Cutting machines

- AS Enterprises – India
- Modern Enterprises – India
- Baumer AG – Germany/USA
- Edge Sweets Company – USA
- Demand Products Inc. – USA
- Wintech Engineering Limited – Australia
- Sunkist Chemical Machinery Limited – Taiwan
- Elitecore Machinery Manufacturing Limited – China

https://doi.org/10.1515/9783110643169-011

Appendix 2 – Recommended solutions for deficiencies in PU parts moulding and block moulding

Parts moulding

Defect	Description	Recommended solutions
Short	Part incomplete	Increase filling volume
Sticking	Part sticking to mould	Increase release coating
Curing cycle too long	Mould opening delayed	Increase mould temperature – check cooling system
Surface patches	White patches	Increase mixing time before pour
Flash	Material seeping out	Reduce filling volume – increase clamping force
Colour patches	Surface colour variations	Check compatibility of dye/pigment in polyol – increase dispersion time

Block moulding

Defect	Description	Recommended solutions
Bottom cavitation	Bottom eaten away	Look for errors in metering – decrease tin catalyst
Dense bottom skin	Hard foam at bottom	Increase silicone level
Creeping cream line	Cream line moves back	Lower amine catalyst level
Smoking	Excessive vapours	Reduce isocyanate level
Tacky bun surface	Foam block surface sticky and too long	Increase catalyst levels – look for errors in metering
Flashing sparklers	Effervescence on rising foam	Decrease isocyanate – decrease amine/silicone
Friable skin	Skin flakes off at touch	Increase/change amine catalyst – increase component temperatures
Gross splits	Vertical/horizontal foam splits	Increase tin catalyst – decrease amine/water content
Heavy skin	Thick high-density skins	Increase all catalysts – increase isocyanate content
Moon craters	Small pockmarks on bun	Reduce air entrapment in pour – minimise splashing on pour
Pee holes	Small spherical holes	Increase silicone – decrease mixing speed
Relaxation	Foam rises and goes down	Check/increase tin catalyst/silicone levels – reduce amine catalyst level – reduce mixing speed – reduce nucleation

Appendix 3 – Glossary of PU terms

Term	Definition
Additive	A material added to promote or alter the final properties but does not take part in the chemical reaction, for example, fillers, pigments and flame retardants.
Blowing agent	An additive to produce a cellular foam. The type used may influence the insulating properties of the final foam.
Blend	A mixture of two or more components, for example, polyols in a foam formulation.
Bun	A large portion cut from a larger block, for example, continuous foaming process.
Casting	Filling of open moulds with liquid polyurethane.
Catalyst	An additive in a PU system which initiates a chemical reaction and also increase or decrease rate of reactions.
Cell	Individual cavities of a foam material. Can be open or closed.
Cell structure	Open cells – cells in a foam with no barriers in between allowing gases or liquid to flow through the foam. Closed cells are enclosed by a continuous membrane without any passageways.
Chain reaction	Lengthening of the main chain or backbone of polymer molecules by end-to-end attachment
Component	A PU formulation is made up of several separate components which are directly metred into the mix head.
Core	Internal portion of a moulded foam which is free from a skin usually used to check density.
Cream time	A measure of the time from liquid to the beginning of the chemical reaction. PUs normally turn into a cream colour.
Curing agent	Additive added to increase rate of cure.
Cycle time	Start to end of a complete moulding cycle measured as time.
Dead time	Time taken for a deformed foam to slowly regain its original shape.
De-mould time	Time from dispensing PU mixture into a mould and removing from mould.
Density	Weight per unit volume expressed as kg/cu.m or lbs/cu.ft.
Dew point	Temperature at which vapour begins to condense.

(continued)

(continued)

Term	Definition
Elongation	The increase in length of a foam specimen before rupture. Expressed as a percentage of the original length.
Exothermic	PU reactions are exothermic meaning that they emit heat.
Filler	Materials such as calcium carbonate, silica or biomass powder to increase load factor and also reduce foam costs.
Flame retardant	Additive used in foam formulations to reduce or retard its tendency to burn.
Friable	Refers to the crumbling or powdering of a foam when surface is rubbed.
Gel time	Time required for a foam to be developed and strengthened to be stable.
Hardness	Property referring to the resistance of indentation.
Hydrolysis	Breakdown of polymers in the presence of water.
Hydroxyl group	The combined oxygen and hydrogen radical (–OH) which forms the reactive group in polyols.
Impact resistance	Ability to withstand mechanical or physical blows without the loss of protective properties.
K-Value	Heat transfer coefficient. Lower the better for insulation.
MDI	Abbreviation for diphenyl methane diisocyanate.
Microcellular	An elastomer of cellular structure having a density between 1.3 and 1.2.
Mix time	Mixing time of a foam mixture before pour/casting into a mould.
Mil	One thousandth of an inch(.001).
Moulded density	Density of a foam when fully cured in its final shape.
Moulding	Process of producing a foam product using a mould.
NDI	Naphthalene diisocyanate.
NCO	Nitrogen/carbon/oxygen. Chemical formula for an isocyanate group.
Open pour	Pouring a foam mixture into an open mould, for example, large block mould – where the foam will rise freely.

(continued)

Term	Definition
Over packing	Dispensing more material than needed to increase density. Too much may cause seepage and result in flash.
Polyisocyanurate (PIR)	Modified type of polyurethane foam which exhibits improved resistance to high temperatures.
Polyester	Polymeric compounds with reactive hydroxyl groups containing ester linkages.
Polyether	Polymeric compounds with reactive groups containing ether linkages.
Polymer	High molecular weight compound, natural or synthetic, whose chemical structure can be represented by repeating small units.
Polyol	Chemical compound with more than one reactive hydroxyl group attached to the molecule.
Post cure	Curing time after removal of foam product from mould. Can be accelerated using elevated temperatures.
PTMEG	Poly tetra methylene ether glycol.
PU	Abbreviation for polyurethane.
Release agent	Material applied to a mould surface for easy release of moulded product.
RIM	Reaction injection moulding. A process where a reacting PU mix is injected into a closed mould.
Rise time	Time period where the foam rises and finally stops
Self-skinning	A foam mixture forming its own skin inside a closed mould.
Skin	A high-density skin formed on the outer surface of a moulded product. This occurs when the surface cools faster than the core.
Slabstock	A polyurethane foam which is made as a continuous block.
Systems	In polyurethane foam manufacture, the combination of components.
Tack-free time	Time period from pour to foam surface losing its stickiness.
TDI	Abbreviation for toluene diisocyanate.
Thermal conductivity	Rate of heat transfer through a thickness of foam of a known area. The lower, the better for insulation purposes.

(continued)

(continued)

Term	Definition
Thermoplastic	A plastic material which can be melted, moulded several times without permanent chemical change, for example, polyethylene and polystyrene.
Thermoset	A plastic material which cannot be re-used after moulding once, for example, PU.
Thixotropic	Property of decreasing viscosity with increasing shear stress but thickens back when stress is removed.
Venting	Controlled release of gas/air from inside a mould through holes/slots.
Viscosity	Measure of the thickness of a liquid. Lower the number, thinner the liquid
Volatile organic components	VOCs are organic materials which evaporate at normal temperatures and pressure.

Glossary: Courtesy of ERA Polymers Ltd. – modified by the author.

Abbreviations

ASTM	American Standards for Testing and Materials
CPS	centipoise
FIFO	first-in, first-out
HR	high resilience
ISO	isocyanate
IFD	indentation force deflection
ISO	International Organization for Standardization
MDI	diphenyl methane diisocyanate
MSDS	material safety data sheets
MW	molecular weight
NOP	natural oil polyol
PUR	polyurethane
SPF	spray polyurethane foam
SPC	statistical process control
TDI	toluene di-isocyanate
PBW	parts by weight
TPU	thermoplastic polyurethane
TPE	thermoplastic elastomer
RIM	reaction injection moulding
VE	viscoelastic

https://doi.org/10.1515/9783110643169-012

Index

https://doi.org/10.1515/9783110643169-013

www.ingramcontent.com/pod-product-compliance
Lightning Source LLC
Chambersburg PA
CBHW081527220326
41598CB00036B/6353